—— PRAISE FOR ——

ARE WE SMART ENOUGH TO KNOW HOW SMART ANIMALS ARE?

WINNER OF THE GOODREADS CHOICE AWARD FOR BEST SCIENCE & TECHNOLOGY

"When I was growing up, aspiring naturalists or behavioural scientists would be given a copy of Konrad Lorenz's zoological investigation *King Solomon's Ring* to inspire them. With its wide-ranging and thought-provoking content, *Are We Smart Enough to Know How Smart Animals Are?* is an appropriate 21st-century replacement. If you are at all interested in what it is to be an animal, human or otherwise, you should read this book."

—THE GUARDIAN

"*Are We Smart Enough to Know How Smart Animals Are?* will completely change your perceptions of the abilities of animals. This book takes the reader on a fascinating journey of discovery into the world of animal problem-solving."

—TEMPLE GRANDIN,
author of *Animals in Translation* and *Animals Make Us Human*

"A thoughtful, balanced argument . . . written simply enough for nonspecialists but with enough detail to engage academics who want a concise review of the field outside their own areas of expertise. The take-home message may be not only whether humans are smart enough to evaluate nonhuman intelligence, but also whether we are humble and open-minded enough to accept that humans may sometimes not be superior to the nonhumans with whom we share the world."

—IRENE PEPPERBERG,
author of *Alex & Me: How a Scientist and a Parrot Discovered a Hidden World of Animal Intelligence— and Formed a Deep Bond in the Process*

More praise for
Are We Smart Enough to Know How Smart Animals Are?

"A beautifully written and delightfully conceived popular science book, written by an eminent researcher who has dedicated his career to making the general public aware of just how smart animals are. *Are We Smart Enough* is on par with de Waal's pioneering book *Chimpanzee Politics*: It is both a lovely read and, given its provocative premise, one that may antagonize the critic."

      —*Science*

"So, are we 'smart enough to know how smart animals are'? The question will occur to you many times as you read Frans de Waal's remarkable distillations of science in this astonishingly broad-spectrum book. I guarantee one thing: readers come away a *lot* smarter. As this book shows, we are here on Planet Earth with plenty of intelligent company."

      —Carl Safina, author of *Beyond Words:*
       What Animals Think and Feel

"Frans de Waal's groundbreaking research has long challenged scientists, philosophers, and theologians to rethink the place of humans in the natural world, showing that we aren't the only species with strategic 'political' behavior, elements of empathy, a sense of justice, and high intelligence. Here he covers not only primates, but a much wider range of species, showing his unique ability to translate the latest findings into sparkling, accessible, provocative books for the thinking public."

      —Robert M. Sapolsky, author of *Why Zebras Don't Get Ulcers*

"You can't help but get a sense that de Waal has placed another nail in the coffin of behaviorism. In animal after animal, de Waal shows the depths of their intelligence and triumphantly affirms that, yes, we are smart enough to see it, and the clues have been there all along."

      —Gregory Berns, author of *How Dogs Love Us*

"Frans de Waal brilliantly demonstrates through scientific evidence, inspiring stories, and common sense that we must fully appreciate the continuous evolutionary process that led to intelligence—understanding situations, reasoning, learning, emotional and empathic knowledge, communication, planning, creativity, and problem solving—and to other amazing cognitive skills that allow various species to best survive, each in their own way. A must for those who aspire to transcend the biases of both anthropocentrism and anthropodenial."
—Matthieu Ricard, author of *Altruism: The Power of Compassion to Change Yourself and the World*

"Engaging and provocative . . . de Waal illuminates the latest ideas and thinking about animal minds and emotions. . . . He challenges us to accept the ultimate findings of this research: Our mental skills are the product of evolution, and all animals from spiders to octopuses to ravens and apes are thinkers in their own ways. And he asks us perhaps the most daunting question of all: Are we really smart enough to understand the minds of other animals?"
—Virginia Morell, author of *Animal Wise: How We Know Animals Think and Feel*

"A thoughtful and easy read, packed with information stemming from detailed empirical research, and one of de Waal's most comparative works that goes well beyond the world of nonhuman primates with whom he's most familiar."
—Marc Bekoff, *Psychology Today*

"The book is not only full of information and thought-provoking, it's also a lot of fun to read."
—Nancy Szokan, *The Washington Post*

"Walks us through research revealing what a wide range of animal species are actually capable of. . . . [I]t all deals a pretty fierce wallop to our sense of specialness."
—Jon Mooallem, *The New York Times Book Review*

"Engaging and informative."
—*The New York Times Book Review*, Editors' Choice

"A good book. Read it instead of watching TV or playing video games. The whole world will be better as a result."
　　　　　—Maria Rodale, *The Huffington Post*

"A fascinating history of the study of animal behavior and cognition."
　　　　　—*Bark*

"Amazing. . . . [T]he clarity of [de Waal's] writing makes for a highly readable book. . . . [A] trip to the zoo may never be the same."
　　　　　—*Kirkus Reviews*, starred review

"Thoroughly engaging, remarkably informative, and deeply insightful. . . . [D]e Waal teaches readers as much about humankind as he does about our nonhuman relatives."
　　　　　—*Publishers Weekly*, boxed and starred review

"This insightful and fascinating work by a scientist who has been at the forefront of new thinking about primates and what it means to be human is highly recommended. De Waal fans and general readers interested in the field of animal cognition will be delighted."
　　　　　—*Library Journal*, boxed and starred review

Also by Frans de Waal

The Bonobo and the Atheist (2013)

The Age of Empathy (2009)

Primates and Philosophers (2006)

Our Inner Ape (2005)

My Family Album (2003)

The Ape and the Sushi Master (2001)

Bonobo (1997)

Good Natured (1996)

Peacemaking among Primates (1989)

Chimpanzee Politics (1982)

ARE WE SMART ENOUGH TO KNOW HOW SMART ANIMALS ARE?

Frans de Waal

With drawings by the author

W. W. Norton & Company
Independent Publishers Since 1923
New York • London

Copyright © 2016 by Frans de Waal

All rights reserved
Printed in the United States of America
First published as a Norton paperback 2017

For information about permission to reproduce selections from this book,
write to Permissions, W. W. Norton & Company, Inc.,
500 Fifth Avenue, New York, NY 10110

For information about special discounts for bulk purchases, please contact
W. W. Norton Special Sales at specialsales@wwnorton.com or 800-233-4830

Manufacturing by Quad Graphics Fairfield
Book design by Dana Sloan
Production manager: Louise Mattarelliano

Library of Congress Cataloging-in-Publication Data

Names: Waal, F. B. M. de (Frans B. M.), 1948– , author.
Title: Are we smart enough to know how smart animals are? / Frans de Waal ;
with drawings by the author.
Description: First edition. | New York : W. W. Norton & Company, 2016. |
   Includes bibliographical references and index.
Identifiers: LCCN 2015049994 | ISBN 9780393246186 (hardcover)
Subjects: LCSH: Animal intelligence. | Psychology, Comparative.
Classification: LCC QL785 .W127 2016 | DDC 591.5/13—dc23
LC record available at http://lccn.loc.gov/2015049994

ISBN 978-0-393-35366-2 pbk.

W. W. Norton & Company, Inc.
500 Fifth Avenue, New York, N.Y. 10110
www.wwnorton.com

W. W. Norton & Company Ltd.
15 Carlisle Street, London W1D 3BS

1 2 3 4 5 6 7 8 9 0

For Catherine,
whom I was smart enough to marry

CONTENTS

ARE WE
SMART ENOUGH
TO KNOW HOW SMART
ANIMALS ARE?

PROLOGUE

The difference in mind between man and the higher animals, great as it is, certainly is one of degree and not of kind.

—Charles Darwin (1871)[1]

One early November morning, while the days were getting colder, I noticed that Franje, a female chimpanzee, was gathering all the straw from her bedroom. She took it under her arm out onto the large island at the Burgers' Zoo, in the Dutch city of Arnhem. Her behavior took me by surprise. First of all, Franje had never done this before, nor had we ever seen other chimps drag straw outside. Second, if her goal was to stay warm during the day, as we suspected, it was notable that she collected the straw while at a cozy temperature inside a heated building. Instead of reacting to the cold, she was bracing for a temperature she could not actually feel. The most reasonable explanation would be that she extrapolated from the previous shivering day to the weather expected today. In any case, later on she stayed nice and warm with little Fons, her son, in the straw nest she'd built.

I never cease wondering about the mental level at which animals operate, even as I know full well that a single story is not enough to draw conclusions. But those stories inspire observations and experiments that do help us sort out what's going on. The science fiction novelist Isaac Asimov reportedly once said, "The most exciting phrase to hear in science, the one that heralds new discoveries, is not 'Eureka!' but 'That's funny.'" I know this thought all too well. We go through a long process of watching our animals, being intrigued and surprised by their actions, systematically testing our ideas about them, and arguing with colleagues over what the data actually mean. As a result, we are rather slow to accept conclusions, and disagreement lurks around every corner. Even if the initial observation is simple (an ape collects a pile of straw), the repercussions can be enormous. The question as to whether animals make plans for the future, as Franje seemed to be doing, is one that science is currently quite preoccupied with. Specialists speak of *mental time travel*, *chronesthesia*, and *autonoesis*, but I will avoid such arcane terminology and try to translate the progress into ordinary language. I will relate stories of the everyday use of animal intelligence as well as offer actual evidence from controlled experiments. The first tells us what purpose cognitive capacities serve, while the second helps us rule out alternative explanations. I value both equally, even though I realize that stories make easier reading than experiments.

Consider the related question as to whether animals say goodbye as well as hello. The latter is not hard to fathom. Greeting is a response to the appearance of a familiar individual after an absence, such as your dog jumping up at you as soon as you walk through the door. Internet videos of soldiers being saluted by pets upon return from abroad suggest a connection between the length of separation and the intensity of the greeting. We can relate to this connection since it applies to us as well. No grand cognitive theories are necessary to account for it. But what about saying goodbye?

We dread to say farewell to someone we love. My mother cried when I moved across the Atlantic, even though we both realized that my absence would not last forever. Saying goodbye presupposes the realization of future separation, which is why it is rare in animals. But here, too, I have a story. I once trained a female chimpanzee named Kuif to bottle-feed an adopted infant of her species. Kuif acted in every way as the infant's mother but lacked sufficient milk of her own to nurse her. We would hand her a bottle of warm milk, which she would carefully give to the baby ape. Kuif got so good at this that she'd even briefly withdraw the bottle if the baby needed to burp. This project required that Kuif and the baby, which she kept on her body day and night, be called inside for a feeding during the daytime while the rest of the colony remained outside. After a while we noticed that instead of coming in right away, Kuif would make a long detour. She'd do the rounds on the island, visiting the alpha male, the alpha female, and several good friends, giving each one a kiss, before she'd walk toward the building. If the others were asleep, she'd wake them up for her goodbyes. Again, the behavior itself was simple, yet the precise circumstances made us wonder about the underlying cognition. Like Franje, Kuif seemed to be thinking ahead.

But what about skeptics who believe that animals are by definition trapped in the present, and only humans contemplate the future? Are they making a reasonable assumption, or are they blinkered as to what animals are capable of? And why is humanity so prone to downplay animal intelligence? We routinely deny them capacities that we take for granted in ourselves. What is behind this? In trying to find out at what mental level other species operate, the real challenge comes not just from the animals themselves but also from within us. Human attitudes, creativity, and imagination are very much part of the story. Before we ask if animals possess a certain kind of intelligence, especially one that we cherish in ourselves, we need to overcome internal resistance to even consider the possibility. Hence this book's central question: "Are we smart enough to know how smart animals are?"

The short answer is "Yes, but you'd never have guessed." For most of the last century, science was overly cautious and skeptical about the intelligence of animals. Attributing intentions and emotions to animals was seen as naïve "folk" nonsense. We, the scientists, knew better! We never went in for any of this "my dog is jealous" stuff, or "my cat knows what she wants," let alone anything more complicated, such as that animals might reflect on the past or feel one another's pain. Students of animal behavior either didn't care about cognition or actively opposed the whole notion. Most didn't want to touch the topic with a ten-foot pole. Fortunately, there were exceptions—and I will make sure to dwell on those, since I love the history of my field—but the two dominant schools of thought viewed animals as either stimulus-response machines out to obtain rewards and avoid punishment or as robots genetically endowed with useful instincts. While each school fought the other and deemed it too narrow, they shared a fundamentally mechanistic outlook: there was no need to worry about the internal lives of animals, and anyone who did was anthropomorphic, romantic, or unscientific.

Did we have to go through this bleak period? In earlier days, the thinking was noticeably more liberal. Charles Darwin wrote extensively about human and animal emotions, and many a scientist in the nineteenth century was eager to find higher intelligence in animals. It remains a mystery why these efforts were temporarily suspended, and why we voluntarily hung a millstone around the neck of biology—which is how the great evolutionist Ernst Mayr characterized the Cartesian view of animals as dumb automatons.[2] But times are changing. Everyone must have noticed the avalanche of knowledge emerging over the last few decades, diffused rapidly over the Internet. Almost every week there is a new finding regarding sophisticated animal cognition, often with compelling videos to back it up. We hear that rats may regret their own decisions, that crows manufacture tools, that octopuses recognize human faces, and that special neurons allow monkeys to learn from each other's mistakes. We speak openly about culture in animals and about their empathy and friendships. Nothing is off limits anymore, not even the rationality that was once considered humanity's trademark.

In all this, we love to compare and contrast animal and human intelligence, taking ourselves as the touchstone. It is good to realize, though, that this is an outdated way of putting it. The comparison is not between humans and animals but between one animal species—ours—and a vast array of others. Even though most of the time I will adopt the "animal" shorthand for the latter, it is undeniable that humans *are* animals. We're not comparing two separate categories of intelligence, therefore, but rather are considering variation within a single one. I look at human cognition as a variety of animal cognition. It is not even clear how special ours is relative to a cognition distributed over eight independently moving arms, each with its own neural supply, or one that enables a flying organism to catch mobile prey by picking up the echoes of its own shrieks.

We obviously attach immense importance to abstract thought and language (a penchant that I am not about to mock while writing a book!), but in the larger scheme of things this is only one way to face the problem of survival. In sheer numbers and biomass, ants and termites may have done a better job than we have, focusing on tight coordination among colony members rather than individual thought. Each society operates like a self-organized mind, albeit one pitter-pattering around on thousands of little feet. There are many ways to process, organize, and spread information, and it is only recently that science has become open-minded enough to treat all these different methods with wonder and amazement rather than dismissal and denial.

So, yes, we are smart enough to appreciate other species, but it has required the steady hammering of our thick skulls with hundreds of facts that were initially poo-pooed by science. How and why we became less anthropocentric and prejudiced is worth reflecting on while considering all that we have learned in the meantime. In going over these developments, I will inevitably inject my own view, which emphasizes evolutionary continuity at the expense of traditional dualisms. Dualisms between body and mind, human and animal, or reason and emotion may sound useful, but they seriously distract from the larger picture. Trained as a biologist and ethologist, I have little patience with the paralyzing skepticism of the past. I doubt that it was worth the oceans of ink that we, myself included, have spent on it.

In writing this book, I do not seek to provide a comprehensive and systematic overview of the field of evolutionary cognition. Readers may find such reviews in other, more technical books.[3] Instead, I will pick and choose from among many discoveries, species, and scientists, so as to convey the excitement of the past twenty years. My own specialty is primate behavior and cognition, an area that has greatly affected others as it has been at the forefront of discovery. Having been part of this field since the 1970s, I have known many of the players firsthand— human as well as animal—which allows me to add a personal touch. There is plenty of history to dwell on. The growth of this field has been an adventure—some would say, a roller-coaster ride—but it remains endlessly fascinating, since behavior is, as the Austrian ethologist Konrad Lorenz put it, the liveliest aspect of all that lives.

I | MAGIC WELLS

What we observe is not nature in itself,
but nature exposed to our method of questioning.

—Werner Heisenberg (1958)[1]

On Becoming a Bug

Opening his eyes, Gregor Samsa woke up inside the body of an unspecified animal. Equipped with a hard exoskeleton, the "horrible vermin" hid under the sofa, crawled up and down walls and ceilings, and loved rotten food. Poor Gregor's transformation inconvenienced and disgusted his family to the point that his death came as a relief.

Franz Kafka's *Metamorphosis*, published in 1915, was an odd opening salvo for a less anthropocentric century. Having selected a repulsive creature for metaphorical effect, the author forced us from the very first page to imagine what it is like to be a bug. At around the same time, Jakob von Uexküll, a German biologist, drew attention to the animal point of view, calling it its *Umwelt*. To illustrate this new concept (German for the "surrounding world"), Uexküll took us on a stroll through various worlds. Each organism senses the environment in its own way, he said. The eyeless tick climbs onto a grass stem to await the smell of butyric acid emanating from mammalian skin. Since experiments have shown that this arachnid can go for eighteen years without food, the tick has ample time to meet a mammal, drop onto her victim, and gorge herself on warm blood. Afterward she is ready to lay her eggs and die. Can we understand the tick's *Umwelt*? It seems incredibly impoverished compared to ours, but Uexküll saw its simplicity as a strength: her goal is well defined, and she encounters few distractions.

Uexküll reviewed other examples, showing that a single environment offers hundreds of realities peculiar to each species. *Umwelt* is quite different from the notion of *ecological niche*, which concerns the habitat that an organism needs for survival. Instead, *Umwelt* stresses an organism's self-centered, subjective world, which represents only a small tranche of all available worlds. According to Uexküll, the various tranches are "not comprehended and never discernible" to all the species that construct them.[2] Some animals perceive ultraviolet light, for example, while others live in a world of smells or, like the star-nosed mole, feel their way around underground. Some sit on the branches of an oak, and others live underneath its bark, while a fox family digs a lair among its roots. Each perceives the same tree differently.

Humans can try to imagine the *Umwelt* of other organisms. Being a highly visual species ourselves, we buy smartphone apps that turn colorful images into those seen by people without color vision. We can walk around blindfolded to simulate the *Umwelt* of the vision-impaired in order to augment our empathy. My most memorable experience with an alien world, however, came from raising jackdaws, small members of the crow family. Two of them flew in and out of my window on the fourth floor of a student dorm, so I could watch their exploits from above. When they were young and inexperienced, I observed them, like any good parent, with great apprehension. We think of flight as something birds do naturally, but it is actually a skill that they have to learn. Landing is the hardest part, and I was always afraid they would crash into a moving car. I began to think like a bird, mapping the environment as if looking for the perfect landing spot, judging a distant object (a branch, a balcony) with this goal in mind. Upon achieving a safe landing, my birds would give happy "caw-caw" calls, after which I would call them to come back, and the whole process would start anew. Once they became expert flyers, I enjoyed their playful tumbling in the wind as if I were flying among them. I entered my birds' *Umwelt*, even though imperfectly.

Whereas Uexküll wanted science to explore and map the *Umwelten* of various species, an idea that deeply inspired students of animal behavior known as ethologists, philosophers of the last century were rather pessimistic. When Thomas Nagel, in 1974, asked, "What is it like to be a bat?" he concluded that we would never know.[3] We have no way of entering the subjective life of another species, he said. Nagel did not seek to know how a human would feel as a bat: he wanted to understand how a bat feels like a bat. This is indeed beyond our comprehension. The same wall between them and us was noted by the Austrian philosopher Ludwig Wittgenstein, when he famously declared, "If a lion could talk, we could not understand him." Some scholars were offended, complaining that Wittgenstein had no idea of the subtleties of animal communication, but the crux of his aphorism was that since our own experiences are so unlike a lion's, we would fail to understand the king of fauna even if he spoke our tongue. In fact, Wittgenstein's reflections extended to people in strange cultures with whom we, even if we know their language, fail to "find our feet."[4] His point was our limited ability to enter the inner lives of others, whether they are foreign humans or different organisms.

Rather than tackle this intractable problem, I will focus on the world that animals live in, and how they navigate its complexity. Even though we can't feel what they feel, we can still try to step outside our own narrow *Umwelt* and apply our imagination to theirs. In fact, Nagel could never have written his incisive reflections had he not heard of the echolocation of bats, which had been discovered only because scientists did try to imagine what it is like to be a bat and did in fact succeed. It is one of the triumphs of our species' thinking outside its perceptual box.

As a student, I listened in amazement as Sven Dijkgraaf, the head of my department at the University of Utrecht, told the story of how, at about my age, he was one of only a handful of people in the world who was able to hear the faint clicks that accompany a bat's ultrasonic vocalizations. The professor had extraordinary hearing. It had been known for more than a century that a blinded bat can still find its way around and safely land on walls and ceilings, whereas a deafened one cannot. A bat without hearing is like a human without sight. No one fully understood how this worked, and bats' abilities were unhelpfully attributed to a "sixth sense." Scientists don't believe in extrasensory perception, however, and Dijkgraaf had to come up with an alternative explanation. Since he could detect a bat's calls, and had noticed that the rate increased when bats encountered obstacles, he suggested that the calls help them traverse their environment. But there was always a tone of regret in his voice about the lack of recognition he had received as the discoverer of echolocation.

This honor had gone to Donald Griffin, and rightly so. Assisted by equipment that could detect sound waves above the 20 kHz range of human hearing, this American ethologist had conducted the ultimate experiments, which furthermore demonstrated that echolocation is more than just a collision warning system. Ultrasound serves to find and pursue prey, from large moths to little flies. Bats possess an astonishingly versatile hunting tool.

No wonder Griffin became an early champion of animal cognition—a term considered an oxymoron until well into the 1980s—because what else is cognition but information processing? *Cognition* is the mental transformation of sensory input into knowledge about the environment and the flexible application of this knowledge. While the term *cognition* refers to the process of doing this, *intelligence* refers more to the ability to do it successfully. The bat works with plenty of sensory input, even if it remains alien to us. Its auditory cortex evaluates sounds bouncing off objects, then uses this information to calculate its distance to the target as well as the target's movement and speed. As if this weren't complex enough, the bat also corrects for its own flight path and distinguishes the echoes of its own vocalizations from those of nearby bats: a form of self-recognition. When insects evolved hearing in order to evade bat detection, some bats responded with "stealth" vocalizations below the hearing level of their prey.

What we have here is a most sophisticated information-processing system backed by a specialized brain that turns echoes into precise perception. Griffin had followed in the footsteps of the pioneering experimentalist Karl von Frisch, who had discovered that honeybees use a waggle dance to communicate distant food locations. Von Frisch once said, "The life of the bee is like a magic well, the more you draw from it, the more there is to draw."[5] Griffin felt the same about echolocation, seeing this capacity as yet another inexhaustible source of mystery and wonder. He called it, too, a magic well.[6]

Since I work with chimpanzees, bonobos, and other primates, people usually don't give me a hard time when I speak of cognition. After all, people are primates, too, and we process our surroundings in similar ways. With our stereoscopic vision, grasping hands, ability to climb and jump, and emotional communication via facial muscles, we inhabit the same *Umwelt* as other primates. Our children play on "monkey bars," and we call imitation "aping," precisely because we recognize these similarities. At the same time, we feel threatened by primates. We laugh hysterically at apes in movies and sitcoms, not because they are inherently funny—there are much funnier-looking animals, such as giraffes and ostriches—but because we like to keep our fellow primates at arm's length. It is similar to how people in neighboring countries, who resemble each other most, joke about each other. The Dutch find nothing to laugh at in the Chinese or the Brazilians, but they relish a good joke about the Belgians.

But why stop at the primates when we are considering cognition? Every species deals flexibly with the environment and develops solutions to the problems it poses. Each one does it differently. We had better use the plural to refer to their capacities, therefore, and speak of intelligence*s* and cognition*s*. This will help us avoid comparing cognition on a single scale modeled after Aristotle's *scala naturae*, which runs from God, the angels, and humans at the top, downward to other mammals, birds, fish, insects, and mollusks at the bottom. Comparisons up and down this vast ladder have been a popular pastime of cognitive science, but I cannot think of a single profound insight it has yielded. All it has done is make us measure animals by human standards, thus ignoring the immense variation in organisms' *Umwelten*. It seems highly unfair to ask if a squirrel can count to ten if counting is not really what a squirrel's life is about. The squirrel is very good at retrieving hidden nuts, though, and some birds are absolute experts. The Clark's nutcracker, in the fall, stores more than twenty thousand pine nuts, in hundreds of different locations distributed over many square miles; then in winter and spring it manages to recover the majority of them.[7]

That we can't compete with squirrels and nutcrackers on this task—I even forget where I parked my car—is irrelevant, since our species does not need this kind of memory for survival the way forest animals braving a freezing winter do. We don't need echolocation to orient ourselves in the dark; nor do we need to correct for the refraction of light between air and water as archerfish do while shooting droplets at insects above the surface. There are lots of wonderful cognitive adaptations out there that we don't have or need. This is why ranking cognition on a single dimension is a pointless exercise. Cognitive evolution is marked by many peaks of specialization. The ecology of each species is key.

The last century has seen ever more attempts to enter the *Umwelt* of other species, reflected in book titles such as *The Herring Gull's World*, *The Soul of the Ape*, *How Monkeys See the World*, *Inside a Dog*, and *Anthill*, in which E. O. Wilson, in his inimitable fashion, offers an ant's-eye view of the social life and epic battles of ants.[8] Following in the footsteps of Kafka and Uexküll, we are trying to get under the skin of other species, trying to understand them on their terms. And the more we succeed, the more we discover a natural landscape dotted with magic wells.

Six Blind Men and the Elephant

Cognition research is more about the possible than the impossible. Nevertheless, the *scala naturae* view has tempted many to conclude that animals lack certain cognitive capacities. We hear abundant claims along the lines of "only humans can do this or that," referring to anything from looking into the future (only humans think ahead) and being concerned for others (only humans care about the well-being of others) to taking a vacation (only humans know leisure time). The last claim once had me, to my own amazement, debating a philosopher in a Dutch newspaper about the difference between a tourist tanning on the beach and a napping elephant seal. The philosopher considered the two to be radically different.

In fact, I find the best and most enduring claims about human exceptionalism to be the funny ones, such as Mark Twain's "Man is the only animal that blushes—or needs to." But, of course, most of these claims are deadly serious and self-congratulatory. The list goes on and on and changes every decade, yet must be treated with suspicion given how hard it is to prove a negative. The credo of experimental science remains that an absence of evidence is not evidence of absence. If we fail to find a capacity in a given species, our first thought ought to be "Did we overlook something?" And the second should be "Did our test fit the species?"

A telling illustration involves gibbons, which were once considered backward primates. Gibbons were presented with problems that

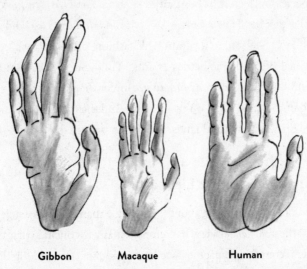

Gibbon   Macaque   Human

The gibbon's hand lacks a fully opposable thumb. It is suited for grasping branches rather than for picking up items from a flat surface. Only when their hand morphology was taken into account did gibbons pass certain intelligence tests. Here a comparison between the hands of a gibbon, a macaque, and a human. After Benjamin Beck (1967).

required them to choose between various cups, strings, and sticks. In test after test, these primates fared poorly compared to other species. Tool use, for example, was tested by dropping a banana outside their cage and placing a stick nearby. All they had to do to get the banana was pick up the stick to move it closer. Chimpanzees will do so without hesitation, as will many manipulative monkeys. But not gibbons. This was bizarre given that gibbons (also known as "lesser apes") belong to the same large-brained family as humans and apes.

In the 1960s an American primatologist, Benjamin Beck, took a fresh approach.[9] Gibbons are exclusively arboreal. Known as *brachiators*, they propel themselves through trees by hanging by their arms and hands. Their hands, which have tiny thumbs and elongated fingers, are specialized for this kind of locomotion: gibbon hands act more like hooks than like the versatile grasping and feeling organs of most other primates.

Beck, realizing that the gibbon's *Umwelt* barely includes the ground level and that its hands make it impossible to pick up objects from a flat surface, redesigned a traditional string-pulling task. Instead of presenting strings lying on a surface, as had been done before, he elevated them to the animal's shoulder level, making them easier to grasp. Without going into detail—the task required the animal to look carefully at how a string was attached to food—the gibbons solved all the problems quickly and efficiently, demonstrating the same intelligence as other apes. Their earlier poor performance had had more to do with the way they were tested than with their mental powers.

Elephants are another good example. For years, scientists believed them incapable of using tools. The pachyderms failed the same out-of-reach banana test, leaving the stick alone. Their failure could not be attributed to an inability to lift objects from a flat surface, because elephants are ground dwellers and pick up items all the time, sometimes tiny ones. Researchers concluded that they just didn't get the problem. It occurred to no one that perhaps we, the investigators, didn't get the elephant. Like the six blind men, we keep turning around and poking the big beast, but we need to remember that, as Werner Heisenberg put it, "what we observe is not nature in itself, but nature exposed to our method of questioning." Heisenberg, a German physicist, made this observation regarding quantum mechanics, but it holds equally true for explorations of the animal mind.

In contrast to the primate's hand, the elephant's grasping organ is also its nose. Elephants use their trunks not only to reach food but also to sniff and touch it. With their unparalleled sense of smell, these animals know exactly what they are going for. But picking up a stick blocks their nasal passages. Even when they bring the stick close to the food, it impedes their feeling and smelling it. It is like sending a blindfolded child out on an Easter egg hunt.

What sort of experiment, then, would do justice to the animal's special anatomy and abilities?

On a visit to the National Zoo in Washington, D.C., I met Preston Foerder and Diana Reiss, who showed me what Kandula, a young elephant bull, can do when the problem is presented differently. The scientists hung fruit high up above Kandula's enclosure, just out of his reach. They gave the elephant several sticks and a sturdy square box. Kandula ignored the sticks but, after a while, began kicking the box with his foot. He kicked it many times in a straight line until it was right underneath

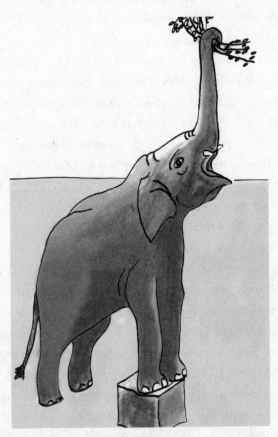

Elephants were believed to be inept tool users based on the assumption that they should use their trunk. In a tool task that bypassed the trunk, however, Kandula had no trouble reaching green branches hanging high above his head. He went out of his way to fetch a box to stand on.

the fruit. He then stood on the box with his front legs, which enabled him to reach the food with his trunk. An elephant, it turns out, can use tools—if they are the right ones.

As Kandula munched his reward, the investigators explained to me how they had varied the setup, making life more difficult for the elephant. They had put the box in a different section of the yard, out of view, so that when Kandula looked up at the tempting food, he would need to recall the solution while distancing himself from his goal to fetch the tool. Apart from a few large-brained species, such as humans, apes, and dolphins, not many animals will do this, but Kandula did it without hesitation, fetching the box from great distances.[10]

Clearly, the scientists had found a species-appropriate test. In search of such methods, even something as simple as size can matter. The largest land animal cannot always be tested with human-sized tools. In one experiment researchers conducted a mirror test—to evaluate whether an animal recognizes its own reflection. They placed a mirror on the floor outside an elephant cage. Measuring only 41 by 95 inches, it was angled up so that the elephant probably mostly saw its legs moving behind two layers of bars (since the mirror doubled them). When the elephant received a body mark that was visible only with assistance of the mirror, it failed to touch it. The verdict was that the species lacked self-awareness.[11]

But Joshua Plotnik, then a student of mine, modified the test. He gave elephants at the Bronx Zoo access to an eight-foot-square mirror placed directly inside their enclosure. They could feel it, smell it, and look behind it. Close-up exploration is a critical step, for apes and humans as well; that had been impossible in the earlier study. In fact, the elephants' curiosity worried us, as the mirror was mounted on a wooden wall that was not designed to support climbing pachyderms. Elephants normally don't stand up against structures, so having a four-ton animal lean on a flimsy wall in order to see and smell what was behind the mounted mirror scared us to death. Clearly, the animals were motivated to find out what the mirror was all about, but if the wall had collapsed, we might have ended up chasing elephants in New York traffic! Fortunately, the wall held, and the animals got used to the mirror.

One Asian elephant, named Happy, recognized her reflection. Marked with a white cross on her forehead above her left eye, she repeatedly rubbed the mark while standing in front of the mirror. She connected her reflection with her own body.[12] By now, years later, Josh has tested many more animals at Think Elephants International, in Thailand, and our conclusion holds: some Asian elephants recognize themselves in the mirror. Whether the same can be said of African elephants is hard to tell, because up to now our experiments have resulted in a lot of destroyed mirrors due to this species' tendency to examine new items with vigorous tusk action. This makes it hard to decide between poor performance and poor equipment. Obviously, the destruction of mirrors is no reason to conclude that African elephants lack mirror self-recognition. We are just dealing with species-typical treatment of novel items.

The challenge is to find tests that fit an animal's temperament, interests, anatomy, and sensory capacities. Faced with negative outcomes, we need to pay close attention to differences in motivation and attention. One cannot expect a great performance on a task that fails to arouse interest. We ran into this problem while studying face recognition in chimpanzees. At the time, science had declared humans unique, since we were so much better at identifying faces than any other primate. No one seemed bothered by the fact that other primates had been tested mostly on human faces rather than those of their own kind. When I asked one of the pioneers in this field why the methodology had never moved beyond the human face, he answered that since humans differ so strikingly from one another, a primate that fails to tell members of our species apart will surely also fail at its own kind.

But when Lisa Parr, one of my coworkers at the Yerkes National

Primate Research Center in Atlanta, tested chimpanzees on photo-graphs of their own species, she found that they excelled at it. Select-ing images on a computer screen, they would see one chimpanzee portrait immediately followed by a pair of others. One portrait of the pair would be a different picture of the same individual as presented before, while the other would show a different individual. Having been trained to detect similarity (a procedure known as matching to sam-ple), the chimpanzees had no trouble recognizing which portrait most resembled the first. The apes even detected family ties. After having seen a female portrait, they were given a choice between two juvenile faces, one of which was the offspring of the female shown before. They picked the latter based purely on physical similarity, since they did not know any of the depicted apes in real life.[13] In much the same way, we can leaf through someone else's family album and quickly notice who are blood relatives and who are in-laws. As it turns out, chimpanzee face recognition is as keen as ours. It is now widely accepted as a shared capacity, especially since it engages the same brain areas in humans and other primates.[14]

In other words, what is salient to us—such as our own facial fea-tures—may not be salient to other species. Animals often know only what they *need* to know. The maestro of observation, Konrad Lorenz, believed that one could not investigate animals effectively without an intuitive understanding grounded in love and respect. He saw such intuitive insight as quite separate from the methodology of the natural sciences. To marry it productively with systematic research is both the challenge and the joy of studying animals. Promoting what he called the *Ganzheitsbetrachtung* (holistic contemplation), Lorenz urged us to grasp the whole animal before zooming in on its various parts.

One cannot master set research tasks if one makes a single part the focus of interest. One must, rather, continuously dart from one part to another—in a way that appears extremely flighty and unscientific to some thinkers who place value on strictly logical sequences—and one's knowledge of each of the parts must advance at the same pace.[15]

The danger of ignoring this advice was amusingly illustrated when a famous study was replicated. In the study, domestic cats were placed in a small cage; they would wander about impatiently meowing—and in the process rub against the cage interior. In so doing, they accidentally moved a latch that opened a door, which allowed them to get out of the cage and eat a scrap of fish nearby. The more trials a cat performed, the quicker she'd escape. The investigators were impressed that all the tested cats showed the same stereotyped rubbing pattern, which they thought they had taught them with food rewards. First developed by

Edward Thorndike's cats were considered to have proven the "law of effect." By rubbing against a latch inside a cage, a cat could open a door and escape, which would gain her a fish. Decades later, however, it was shown that the cats' behavior had nothing to do with the prospect of reward. The animals escaped just as well without the fish. The presence of friendly people was all that was needed to elicit the flank rubbing that marks all feline greeting behavior. After Thorndike (1898).

Edward Thorndike in 1898, this experiment was considered proof that even seemingly intelligent behavior (such as escaping from a cage) can be fully explained by trial-and-error learning. It was a triumph of the "law of effect," according to which behavior with pleasant consequences is likely to be repeated.[16]

When the American psychologists Bruce Moore and Susan Stuttard replicated this study decades later, however, they found that the cats' behavior was nothing special. The cats performed the usual *Köpfchengeben* (German for "head giving") that all felines—from house cats to tigers—use in greeting and courting. They rub their head or flank against the object of affection or, if the object of affection is inaccessible, redirect the rubbing to inanimate objects, such as the legs of a kitchen table. The investigators showed that the food reward was not needed: the only meaningful factor was the presence of friendly people. Without training, every caged cat that saw a human observer rubbed its head, flank, and tail against the latch and got out of the cage. Left alone, however, the cats were unable to escape, since they never performed any rubbing.[17] Instead of a learning experiment, the classical study had been a greeting experiment! The replication was published under the telling subtitle "Tripping over the Cat."

The lesson is that before scientists test any animal, they need to know its typical behavior. The power of conditioning is not in doubt, but the early investigators had totally overlooked a crucial piece of information. They had not, as recommended by Lorenz, considered the whole organism. Animals show many unconditioned responses, or behavior that develops naturally in all members of their species. Reward and punishment may affect such behavior but cannot take credit creating it. The reason all cats responded in the same way derived from natural feline communication rather than operant conditioning.

The field of evolutionary cognition requires us to consider every species in full. Whether we are studying hand anatomy, trunk multifunctionality, face perception, or greeting rituals, we need to familiarize ourselves with all facets of the animal and its natural history before trying to figure out its mental level. And instead of testing animals on abilities that *we* are particularly good at—our own species' magic wells, such as language—why not test them on *their* specialized skills? In doing so, we will not just flatten Aristotle's scale of nature: we will transform it into a bush with many branches. This change in perspective is now feeding the long-overdue recognition that intelligent life is not something we must seek at great expense only in the outer reaches of space. It is abundant here on earth, right underneath our nonprehensile noses.[18]

Anthropodenial

The ancient Greeks believed that the center of the universe was right where they lived. What better place, therefore, than Greece for modern scholars to ponder humanity's place in the cosmos? On a sunny day in 1996, an international group of academics visited the *omphalos* (navel) of the world—a large stone shaped like a beehive—amid the temple ruins on Mount Parnassus. I couldn't resist patting it like a long-lost friend. Right next to me stood "batman" Don Griffin, the discoverer of echolocation and author of *The Question of Animal Awareness*, in which he lamented the misperception that everything in the world turns around us and that we are the only conscious beings.[19]

Ironically, a major theme of our workshop was the anthropic principle, according to which the universe is a purposeful creation uniquely suited for intelligent life, meaning us.[20] At times the discourse of the anthropic philosophers sounded as if they thought the world was made for us rather than the other way around. Planet Earth is at exactly the right distance from the sun to create the right temperature for human life, and its atmosphere has the ideal oxygen level. How convenient! Instead of seeing purpose in this situation, however, any biologist will turn the causal connection around and note that our species is finely adapted to the planet's circumstances, which explains why they are perfect for us. Deep ocean vents are an optimal environment for bacteria thriving on their superhot sulfuric output, but no one assumes that these vents were created to serve thermophile bacteria; rather, we understand that natural selection has shaped bacteria able to live near them.

The backward logic of these philosophers reminded me of a creationist I once saw peel a banana on television while explaining that this fruit is curved in such a way that it conveniently angles toward the human mouth when we hold it in our hand. It also fits perfectly in our mouth. Obviously, he felt that God had given the banana its human-friendly shape, while forgetting that he was holding a domesticated fruit, cultivated for human consumption.

During some of these discussions, Don Griffin and I watched barn swallows flying back and forth outside the conference room window carrying mouthfuls of mud for their nests. Griffin was at least three decades my senior and had impressive knowledge, offering the Latin name of the birds and describing details of their incubation period. At the workshop, he presented his view on consciousness: that it has to be part and parcel of all cognitive processes, including those of animals. My own position is slightly different in that I prefer not to make any firm statements about something as poorly defined as consciousness. No one seems to know what it is. But for the same reason, I hasten to add, I'd never deny it to any species. For all I know, a frog may be conscious. Griffin took a more positive stance, saying that since intentional, intelligent actions are observable in many animals, and since in our own species they go together with awareness, it is reasonable to assume similar mental states in other species.

That such a highly respected and accomplished scientist made this claim had a hugely liberating effect. Even though Griffin was slammed for making statements that he could not back up with data, many critics missed the point, which was that the assumption that animals are "dumb," in the sense that they lack conscious minds, is only that: an assumption. It is far more logical to assume continuity in every domain,

Ape gestures are homologous with those of humans. Not only do they look strikingly human, they occur in roughly similar contexts. Here a female chimpanzee (right) kisses a grizzled alpha male on the mouth during a reconciliation after a fight between them.

Griffin said, echoing Charles Darwin's well-known observation that the mental difference between humans and other animals is one of degree rather than kind.

It was an honor to get to know this kindred spirit and to make my own case regarding anthropomorphism, another theme at the conference. Greek for "human form," the word *anthropomorphism* came about when Xenophanes, in 570 B.C., objected to Homer's poetry because it described the gods as if they looked like people. Xenophanes ridiculed the arrogance behind this assumption—why couldn't they look like horses? But gods are gods, far removed from the present-day liberal use of the word *anthropomorphism* as an epithet to vilify any and all human-animal comparisons, even the most cautious ones.

In my opinion, anthropomorphism is problematic only when the human-animal comparison is a stretch, such as with regards to species distant from us. The fish known as kissing gouramis, for example, don't really kiss in the same way and for the same reasons that humans do. Adult fish sometimes lock their protruding mouths together to settle disputes. Clearly, to label this habit "kissing" is misleading. Apes, on the other hand, do greet each other after a separation by placing their lips gently on each other's mouth or shoulder and hence kiss in a way and under circumstances that greatly resemble human kissing. Bonobos go even further: when a zookeeper familiar with chimpanzees once naïvely accepted a bonobo kiss, not knowing this species, he was taken aback by the amount of tongue that went into it!

Another example: when young apes are being tickled, they make breathy sounds with a rhythm of inhalation and exhalation that resembles human laughter. One cannot simply dismiss the term *laughter* for this behavior as too anthropomorphic (as some have done), because not only do the apes sound like human children being tickled, they show the same ambivalence about it as children do. I have often noticed it myself. They try to push my tickling fingers away, but then come back begging for more, holding their breath while awaiting the next poke in their belly. In this case, I am all for shifting the burden of proof and ask those who wish to avoid humanlike terminology to first prove that a tickled ape, who almost chokes on its hoarse giggles, is in fact in a different state of mind from a tickled human child. Absent such evidence, *laughter* strikes me as the best label for both.[21]

Needing a new term to make my point, I invented *anthropodenial*, which is the a priori rejection of humanlike traits in other animals or animallike traits in us. Anthropomorphism and anthropodenial have an inverse relationship: the closer another species is to us, the more anthropomorphism will assist our understanding of this species and the greater will be the danger of anthropodenial.[22] Conversely, the more distant a species is from us, the greater the risk that anthropomorphism will propose questionable similarities that have come about independently. Saying that ants have "queens," "soldiers," and "slaves" is mere anthropomorphic shorthand. We should attach no more significance to it than we do when we name a hurricane after a person or curse our computer as if it had free will.

The key point is that anthropomorphism is not always as problematic as people think. To rail against it for the sake of scientific objectivity often hides a pre-Darwinian mindset, one uncomfortable with the notion of humans as animals. When we are considering species like the apes, which are aptly known as "anthropoids" (humanlike), however, anthropomorphism is in fact a logical choice. Dubbing an ape's kiss "mouth-to-mouth contact" so as to avoid anthropomorphism deliberately obfuscates the meaning of the behavior. It would be like assigning Earth's gravity a different name than the moon's, just because we think Earth is special. Unjustified linguistic barriers fragment the unity with which nature presents us. Apes and humans did not have enough time to independently evolve strikingly similar behavior, such as lip contact in greeting or noisy breathing in response to tickling. Our terminology should honor the obvious evolutionary connections.

On the other hand, anthropomorphism would be a rather empty exercise if all it did was paste human labels onto animal behavior. The American biologist and herpetologist Gordon Burghardt has called for a *critical anthropomorphism*, in which we use human intuition and knowledge of an animal's natural history to formulate research questions.[23] Thus, saying that animals "plan" for the future or "reconcile" after fights is more than anthropomorphic language: these terms propose testable ideas. If primates are capable of planning, for example, they should hold on to a tool that they can use only in the future. And if primates reconcile after fights, we should see a reduction of tensions as well as improved social relationships after opponents have made up by means of friendly contact. These obvious predictions have by now been borne out by actual experiments and observations.[24] Serving as a means rather than an end, critical anthropomorphism is a valuable source of hypotheses.

Griffin's proposal to take animal cognition seriously led to a new label for this field: *cognitive ethology*. It is a great label, but then I am an ethologist and know exactly what he meant. Unfortunately, the term *ethology* has not universally caught on, and spell-checkers still regularly change it to *ethnology*, *etiology*, or even *theology*. No wonder many ethologists nowadays call themselves behavioral biologists. Other existing labels for cognitive ethology are *animal cognition* and *comparative cognition*. But those two terms have drawbacks, too. *Animal cognition* fails to include humans, so it unintentionally perpetuates the idea of a gap between humans and other animals. The *comparative* label, on the other hand, remains agnostic about how and why we make comparisons. It hints at no framework whatsoever to interpret similarities and differences, least of all an evolutionary one. Even within this discipline, there have been complaints about its lack of theory as well as its habit of dividing animals into "higher" and "lower" forms.[25] The label derives from *comparative psychology,* the name of a field that traditionally has viewed animals as mere stand-ins for humans: a monkey is a simplified human, a rat a simplified monkey, and so on. Since associative learning was thought to explain behavior across all species, one of the field's founders, B. F. Skinner, felt that it hardly mattered what kind of animal one worked on.[26] To prove his point, he entitled a book entirely devoted to albino rats and pigeons *The Behavior of Organisms.*

For these reasons, Lorenz once joked that there was nothing comparative about comparative psychology. He knew what he was talking about, having just published a seminal study on the courtship patterns of twenty different duck species.[27] His sensitivity to the minutest differences between species was quite the opposite of the way comparative psychologists lump animals together as "nonhuman models of human behavior." Think for a second about this terminology, which remains so entrenched in psychology that no one takes notice anymore. Its first implication, of course, is that the only reason to study animals is to learn about ourselves. Second, it ignores that every species is uniquely adapted to its own ecology, because otherwise how could one serve as a model for another? Even the term *nonhuman* grates on me, since it lumps millions of species together by an absence, as if they were missing something. Poor things, they are nonhuman! When students embrace this jargon in their writing, I cannot resist sarcastic corrections in the margin saying that for completeness's sake, they should add that the animals they are talking about are also nonpenguin, nonhyena, and a whole lot more.

Even though comparative psychology is changing for the better, I'd rather avoid its leaden baggage and propose to call the new field *evolutionary cognition*, which is the study of all cognition (human and animal) from an evolutionary standpoint. Which species we study obviously matters a great deal, and humans are not necessarily central to every comparison. The field includes phylogeny, when we trace traits across the evolutionary tree to determine whether similarities are due to common descent, the way Lorenz had done so beautifully for waterfowl. We also ask how cognition has been shaped to serve survival. The agenda of this field is precisely what Griffin and Uexküll had in mind, in that it seeks to place the study of cognition on a less anthropocentric footing. Uexküll urged us to look at the world from the animal's standpoint, saying that this is the only way to fully appreciate animal intelligence.

A century later we are ready to listen.

2 | A TALE OF TWO SCHOOLS

Do Dogs Desire?

Given the prominent role that jackdaws and little silvery fish known as three-spined sticklebacks—my favorite childhood animals—played in the early years of ethology, the discipline was an easy sell to me. I learned about it when, as a biology student, I heard a professor explain the zigzag dance of the stickleback. I was floored: not by what these little fish did but by how seriously science took what they did. It was the first time I realized that what I liked doing best—watching animals behave—could be a profession. As a boy, I had spent hours observing self-caught aquatic life that I kept in buckets and tanks in our backyard. The high point had been breeding sticklebacks and releasing the young back into the ditch from which their parents had come.

Ethology is the biological study of animal behavior that arose in continental Europe right before and after World War II. It reached the English-speaking world when one of its founders, Niko Tinbergen, moved across the Channel. A Dutch zoologist, Tinbergen started out in Leiden and accepted a position in Oxford in 1949. He described the male stickleback's zigzag dance in great detail, explaining how it draws the female to the nest where the male fertilizes her eggs. The male then chases her off and protects the eggs, fanning and aerating them, until they hatch. I had seen it all with my own eyes in an abandoned aquarium—its luxurious algae growth was exactly what the fish needed—including the stunning transformation of silvery males into brightly red and blue show-offs. Tinbergen had noticed that males in tanks in the window-sill of his lab in Leiden would get agitated every time a red mail truck drove by in the street below. Using fish models to trigger courtship and aggression, he confirmed the critical role of a red signal.

Clearly, ethology was the direction I wanted to go in, but before pursuing this goal, I was briefly diverted by its rival discipline. I worked in the lab of a psychology professor trained in the *behaviorist* tradition that dominated comparative psychology for most of the last century. This school was chiefly American but evidently had reached my university in the Netherlands. I still remember this professor's classes, in which he made fun of anyone who believed to know what animals "want," "like," or "feel," carefully neutralizing such terminology with quotation marks. If your dog drops a tennis ball in front of you and looks up at you with wagging tail, do you think she wants to play? How naïve! Who says dogs have desires and intentions? Her behavior is the product of the law of effect: she must have been rewarded for it in the past. The dog's mind, if such a thing even exists, remains a black box.

Its focus on nothing but behavior is what gave behaviorism its name, but I had trouble with the idea that animal behavior could be reduced to a history of incentives. It presented animals as passive, whereas I view them as seeking, wanting, and striving. True, their behavior changes based on its consequences, but they never act randomly or accidentally to begin with. Let's take the dog and her ball. Throw a ball at a puppy, and she will go after it like an eager predator. The more she learns about prey and their escape tactics—or about you and your fake throws— the better a hunter or fetcher she will become. But still, at the root of everything is her immense enthusiasm for the pursuit, which takes her through shrubs, into water, and sometimes through glass doors. This enthusiasm manifests itself before any skill development.

Now, compare this behavior with that of your pet rabbit. It doesn't matter how many balls you throw at him, none of the same learning will take place. Absent a hunting instinct, what is there to acquire? Even if you were to offer your rabbit a juicy carrot for every retrieved ball, you'd be in for a long, tedious training program that would never generate the excitement for small moving objects known of cats and dogs. Behaviorists totally overlooked these natural proclivities, forgetting that by flapping their wings, digging holes, manipulating sticks, gnawing wood, climbing trees, and so on, every species sets up its own learning opportunities. Many animals are driven to learn the things they need to know or do, the way kid goats practice head butts or human toddlers have an insuppressible urge to stand up and walk. This holds even for animals in a sterile box. It is no accident that rats are trained to press bars with their paws, pigeons to peck keys with their beaks, and cats to rub their flanks against a latch. Operant conditioning tends to reinforce what is already there. Instead of being the omnipotent creator of behavior, it is its humble servant.

One of the first illustrations came from the work on kittiwakes by Esther Cullen, a postdoctoral student of Tinbergen. Kittiwakes are seabirds of the gull family; they differ from other gulls in that to deter predators, they nest on narrow cliffs. These birds rarely give alarm calls and do not vigorously defend their nests—they don't need to. But what is most intriguing is that kittiwakes fail to recognize their young. Ground-nesting gulls, in which the young move around after hatching, recognize their offspring within days and do not hesitate to kick out strange ones that scientists place in their nests. Kittiwakes, on the other hand, can't tell the difference between their own and strange young, treating the latter like their own. Not that they need to worry about this situation: fledglings normally stay put at the parental nest. This is, of course, precisely why biologists think kittiwakes lack individual recognition.[1]

For the behaviorist, though, such findings are thoroughly puzzling. Two similar birds differing so starkly in what they learn makes no sense, because learning is supposedly universal. Behaviorism ignores ecology and has little room for learning that is adapted to the specific needs of each organism. It has even less room for an absence of learning, as in the kittiwake, or other biological variation, such as differences between the sexes. In some species, for example, males roam a large area in search of mates, whereas females occupy smaller home ranges. Under such conditions, males are expected to have superior spatial abilities. They need to remember when and where they ran into a member of the opposite sex. Giant panda males travel far and wide through the wet bamboo forest, which is uniformly green in all directions. It is crucial for them to be at the right place at the right time given that females ovulate only once per year and are receptive for just a couple of days—which is why zoos have such trouble breeding this magnificent bear. That males have better spatial abilities than females was confirmed when Bonnie Perdue, an American psychologist, tested pandas at the Chengdu Research Base of Giant Panda Breeding in China. She did so by spreading out food boxes over an outdoor area. Panda males were much better than females at remembering which boxes had recently been baited. In contrast, when the Asian small-clawed otter, a member of the same *arctoidea* (bear-like) family of carnivores, was tested on a similar task, both sexes performed the same. This otter being monogamous, males and females occupy the same territory. Similarly, males of sexually promiscuous rodent species navigate mazes more easily than females, whereas monogamous rodents show no sex difference.[2]

If learning talents are a product of natural history and mating strategies, the whole notion of universality begins to fall apart. We can expect huge variation. Evidence for inborn learning specializations has been steadily mounting.[3] There are many different types, from the way ducklings imprint on the first moving object they see—whether it is their mother or a bearded zoologist—to the song learning of birds and whales and the way primates copy one another's tool use. The more variation we discover, the shakier gets the claim that all learning is essentially the same.[4]

Yet during my student days, behaviorism still ruled supreme, at least in psychology. Luckily for me, the professor's pipe-smoking associate, Paul Timmermans, regularly took me aside to induce some much-needed reflection on the indoctrination I was being subjected to. We worked with two young chimpanzees who offered me my first contact with primates apart from my own species. It was love at first sight. I had never met animals that so clearly possessed a mind of their own. Between puffs of smoke, Paul would ask rhetorically, with a twinkle in his eyes, "Do you really think chimps lack emotions?" He would do so just after the apes had thrown a shrieking temper tantrum for not getting their way, or laughed their hoarse chuckles during roughhousing. Paul would also mischievously ask my opinion about other taboo topics, without necessarily saying that the professor was wrong. One night the chimps escaped and ran through the building, only to return to their cage, carefully closing its door behind them before going to sleep. In the morning, we found them curled up in their straw nests and would not have suspected a thing had it not been for the smelly droppings discovered in the hallway by a secretary. "Is it possible that apes think ahead?" Paul asked when I wondered why the apes had closed their own door. How to deal with such crafty, volatile characters without assuming intentions and emotions?

To drive this point home more bluntly, imagine that you wish to enter a testing room with chimpanzees, as I did every day. I would suggest that rather than rely on some behavioral coding scheme that denies intentionality, you pay close attention to their moods and emotions, reading them the way you would any person's, and beware of their tricks. Otherwise, you might end up like one of my fellow students. Despite the advice we gave him of how to dress for the occasion, he came to his first encounter in a suit and tie. He was sure he could handle such relatively small animals, while mentioning how good he was with dogs. The two chimps were mere juveniles, only four and five years old at the time. But of course, they were already stronger than any grown man, and ten times more cunning than a dog. I still remember the student staggering out of the testing room, having trouble shedding both apes clinging to his legs. His jacket was in tatters, with both sleeves torn off. He was fortunate that the apes never discovered the choking function of his tie.

One thing I learned in this lab was that superior intelligence doesn't imply better test outcomes. We presented both rhesus monkeys and chimpanzees with a simple task, known as haptic (touch) discrimination. They were to stick their hand through a hole to feel the difference between two shapes and pick the correct one. Our goal was to do hundreds of trials per session, but whereas this worked well with the monkeys, the chimps had other ideas. They would do fine on the first dozen trials, showing that the discrimination posed no problem, but then their attention would wander. They'd thrust their hands farther so as to reach me, pulling at my clothes, making laughing faces, banging on the window that separated us, and trying to engage me in play. Jumping up and down, they'd even gesture to the door, as if I didn't know how to get to their side. Sometimes, unprofessionally, I would give in and have fun with them. Needless to say, the apes' performance on the task was well below that of the monkeys, not due to an intellectual deficit but because they were bored out of their minds.

The task was just not up to their intellectual level.

The Hunger Games

Are we open-minded enough to assume that other species have a mental life? Are we creative enough to investigate it? Can we tease apart the roles of attention, motivation, and cognition? Those three are involved in everything animals do; hence poor performance can be explained by any one of them. With the above two playful apes, I opted for tedium to explain their underperformance, but how to be sure? It takes human ingenuity indeed to know how smart an animal is.

It also takes respect. If we test animals under duress, what can we expect? Would anyone test the memory of human children by throwing them into a swimming pool to see if they remember where to get out? Yet the Morris Water Maze is a standard memory test used every day in hundreds of laboratories that make rats frantically swim in a water tank with high walls until they come upon a submerged platform that saves them. In subsequent trials, the rats need to remember the platform's location. There is also the Columbia Obstruction Method, in which animals have to cross an electrified grid after varying periods of deprivation, so researchers can see if their drive to reach food or a mate (or for mother rats, their pups) exceeds the fear of a painful shock. Stress is, in fact, a major testing tool. Many labs keep their animals at 85 percent of typical body weight to ensure food motivation. We have woefully little data on how hunger affects their cognition, although I do remember a paper entitled "Too Hungry to Learn?" about food-deprived chickens that were not particularly good at noticing the finer distinctions of a maze task.[5]

The assumption that an empty stomach improves learning is curious. Think about your own life: absorbing the layout of a city, getting to know new friends, learning to play the piano or do your job. Does food play much of a role? No one has ever proposed permanent food deprivation for university students. Why would it be any different for animals? Harry Harlow, a well-known American primatologist, was an early critic of the hunger reduction model. He argued that intelligent animals learn mostly through curiosity and free exploration, both of which are likely killed by a narrow fixation on food. He poked fun at the Skinner box, seeing it as a splendid instrument to demonstrate the effectiveness of food rewards but not to study complex behavior. Harlow added this sarcastic gem: "I am not for one moment disparaging the value of the rat as a subject for psychological investigation; there is very little wrong with the rat that cannot be overcome by the education of the experimenters."[6]

I was amazed to learn that the nearly century-old Yerkes Primate Center went through an early period in which it tested food deprivation on chimpanzees. In the early years, the center was still located in Orange Park, Florida, before it moved to Atlanta, where it became a major institute for biomedical and behavioral neuroscience research. While still in Florida, in 1955, the center set up an operant conditioning program modeled on procedures with rats, including a drastic reduction in body weight and the replacement of chimp names with numbers. Treating apes as rats proved no success, however. Due to the gigantic tensions this program engendered, it lasted only two years. The director and most of the staff deplored the fasting imposed on their apes and constantly argued with the hard-nosed behaviorists who claimed that this was the only way to give the apes "purpose in life," as they blithely called it. Expressing no interest in cognition—the existence of which they didn't even acknowledge—they investigated reinforcement schedules and the punitive effect of time-outs. Rumor had it that staff sabotaged their project by secretly feeding the apes at night. Feeling unwelcome and unappreciated, the behaviorists left because, as Skinner later put it, "tender-hearted colleagues frustrated [their] efforts to reduce chimpanzees to a satisfactory state of deprivation."[7] Nowadays, we would recognize the friction as about not just methodology but also ethics. That creating morose, grumpy apes through starvation was unnecessary was clear from one of the behaviorists' own attempts with an alternative incentive. Chimpanzee number 141, as he called him, successfully learned a task after each correct choice was rewarded with an opportunity to groom the experimenter's arm.[8]

The difference between behaviorism and ethology has always been one of human-controlled versus natural behavior. Behaviorists sought to dictate behavior by placing animals in barren environments in which they could do little else than what the experimenter wanted. If they didn't, their behavior was classified as "misbehavior." Raccoons, for example, are almost impossible to train to drop coins into a box, because they prefer to hold on to them and frantically rub them together—a perfectly normal foraging behavior for this species.[9] Skinner had no eye for such natural proclivities, however, and preferred a language of control and domination. He spoke of behavioral engineering and manipulation, and not just in relation to animals. Later in life he sought to turn humans into happy, productive, and "maximally effective" citizens.[10] While there is no doubt that operant conditioning is a solid and valuable idea and a powerful modifier of behavior, behaviorism's big mistake was to declare it the only game in town.

Ethologists, on the other hand, are more interested in spontaneous behavior. The first ones were eighteenth-century Frenchmen, who already used the label *ethology*, derived from the Greek *ethos*, "character," to refer to the study of species-typical characteristics. In 1902 the great American naturalist William Morton Wheeler made the English term popular as the study of "habits and instincts."[11] Ethologists did conduct experiments and were not averse to working with captive animals, but still a world of difference lay between Lorenz calling his jackdaws down from the sky or being followed by a gaggle of waddling goslings and Skinner standing before rows of cages with singly housed pigeons, firmly closing his hand around the wings of one of his birds.

Ethology developed its own specialized language about instincts, fixed action patterns (a species' stereotypical behavior, such as the dog's tail wagging), innate releasers (stimuli that elicit specific behavior, such as the red dot on a gull's bill that triggers pecking by hungry chicks), displacement activities (seemingly irrelevant actions resulting from conflicting tendencies, such as scratching oneself before a decision), and so on. Without going into the details of its classical framework, ethology's focus was on behavior that develops naturally in all members of a given species. A central question was what purpose a behavior might serve. Initially, the grand architect of ethology was Lorenz, but after he and Tinbergen met in 1936, the latter became the one to fine-tune the ideas and develop critical tests. Tinbergen was the more analytical and empirical of the two, with an excellent eye for the questions behind observable behavior; he conducted field experiments on digger wasps, sticklebacks, and gulls to pinpoint behavioral functions.[12]

The two men formed a complementary relationship and friendship, which was tested by World War II in which they were on opposite sides. Lorenz served as medical officer in the German army and opportunistically sympathized with Nazi doctrine; Tinbergen was imprisoned for two years by the German occupiers of the Netherlands for joining a protest against the way his Jewish colleagues at the university were treated. Remarkably, both scientists patched things up after the war for the sake of their shared love of animal behavior. Lorenz was the charismatic, flamboyant thinker—he didn't conduct a single statistical analysis in his life—while Tinbergen did the nitty-gritty of actual data collection. I have seen both men speak and can attest to the difference. Tinbergen came across as academic, dry, and thoughtful, whereas Lorenz enthralled his audiences with his enthusiasm and intimate animal knowledge. Desmond Morris, a Tinbergen student famous for writing *The Naked Ape* and other popular books, got his socks knocked off by Lorenz, saying that the Austrian understood animals better than anyone he'd ever met. He described Lorenz's 1951 lecture at Bristol University as follows:

> To describe his performance as a tour de force is an understatement. Looking like a cross between God and Stalin, his presence was overpowering. "Contrary to your Shakespeare," he boomed, "there is madness in my method." And indeed there was. Almost all his discoveries were made by accident and his life consisted largely of a series of disasters with the menageries of animals with which he surrounded himself. His understanding

*of animal communication and display patterns was revelatory. When he
spoke about fish, his hands became fins, when he talked about wolves his
eyes were those of a predator, and when he told tales about his geese his arms
became wings tucked into his sides. He wasn't anthropomorphic, he was the
opposite*—theriomorphic—*he became the animal he was describing.*[13]

A journalist once recounted how she had been sent into Lorenz's office by a receptionist with the words that he was expecting her. His office turned out to be empty. When she asked around, people assured her that he had never left. After a while, she discovered the Nobelist partially submerged in an enormous aquarium built into the office wall. This is how we like our ethologists: as close to their animals as possible. It reminds me of my own encounter with Gerard Baerends, the silverback of Dutch ethology and the very first student of Tinbergen. After my stint in the behaviorist lab, I sought to enter Baerends's ethology course at the University of Groningen to work with the jackdaw colony that flew around the institution's nest boxes. Everyone warned me that Baerends was very strict and did not let just anybody in. When I walked into his office, my eyes were immediately drawn to a large well-kept tank with convict cichlids. Being an avid aquarist myself, I hardly took the time to introduce myself before we launched into a discussion of how these fish raise and guard their fry, which they do extraordinarily well. Baerends must have taken my passion as a good sign, because I was admitted without a problem.

The great novelty of ethology was to bring the perspective of morphology and anatomy to bear on behavior. This was a natural step, because whereas behaviorists were mostly psychologists, ethologists were mostly zoologists. They discovered that behavior is not nearly as fluid and hard to define as it might seem. It has a structure, which can be quite stereotypical, such as the way young birds flutter their wings while begging for food with gaping mouths, or how some fish keep fertilized eggs in their mouth until they hatch. Species-typical behavior is as recognizable

Konrad Lorenz and other ethologists wanted to know how animals behave of their own accord and how it suits their ecology. In order to understand the parent-offspring bond in waterfowl, Lorenz let goslings imprint on himself. They followed the pipe-smoking zoologist around wherever he went.

and measurable as any physical trait. Given their invariant structure and meaning, human facial expressions are another good example. The reason we now have software that reliably recognizes human expressions is that all members of our species contract the same facial muscles under similar emotional circumstances.

Insofar as behavior patterns are innate, Lorenz argued, they must be subject to the same rules of natural selection as physical traits and be traceable from species to species across the phylogenetic tree. This is as true for the mouth brooding of certain fish as it is for primate facial expressions. Given that the facial musculature of humans and chimpanzees is nearly identical, the laughing, grinning, and pouting of both species likely goes back to a common ancestor.[14] Recognition of this parallel between anatomy and behavior was a great leap forward, which is nowadays taken for granted. We all now believe in behavioral evolution, which makes us Lorenzians. Tinbergen's role was, as he put it himself, to act as the "conscience" of the new discipline by pushing for more precise formulations of its theories and developing ways to test them. He was overly modest saying so, though, because in the end it was he who best spelled out the ethological agenda and turned the field into a respectable science.

Keeping It Simple

Despite the differences between ethology and behaviorism, the two schools had one thing in common. Both were reactions against the overinterpretation of animal intelligence. They were skeptical of "folk" explanations and dismissed anecdotal reports. Behaviorism was the more vehement in its rejection, saying that behavior is all we have to go by and that internal processes can be safely ignored. There is even a joke about its complete reliance on external cues, in which one behaviorist asks another after lovemaking: "That was great for you. How was it for me?"

In the nineteenth century, it was perfectly acceptable to talk about the mental and emotional lives of animals. Charles Darwin himself had written a whole tome about the parallels between human and animal emotional expressions. But while Darwin was a careful scientist who double-checked his sources and conducted observations of his own, others went overboard, almost as in a contest of who could come up with the wildest claim. When Darwin chose the Canadian-born George Romanes as his protégé and successor, the stage was set for an avalanche of misinformation. About half the animal stories collected by Romanes sound plausible enough, but others are embellished or plainly unlikely. They range from a story about rats forming a supply line to their hole in the wall, carefully handing down stolen eggs with their forepaws, to one about a monkey hit by a hunter's bullet who smeared his hand with his own blood and held it out to the hunter to make him feel guilty.[15]

Romanes knew the mental operations required for such behavior, he said, by extrapolating from his own. The weakness of his introspective approach was, of course, its reliance on one-time events and on trust in one's own private experiences. I have nothing against anecdotes, especially if they have been caught on camera or come from reputable observers who know their animals; but I do view them as a starting point of research, never an end point. For those who disparage anecdotes altogether, it is good to keep in mind that almost all interesting work on animal behavior has begun with a description of a striking or puzzling event. Anecdotes hint at what is possible and challenge our thinking.

But we cannot exclude that the event was a fluke, never to be repeated again, or that some decisive aspect went unnoticed. The observer may also unconsciously have filled in missing details based on his or her assumptions. These issues are not easily resolved by collecting more anecdotes. "The plural of anecdote is not data," as the saying goes. It is ironic, therefore, that when it was his own turn to find a protégé and successor, Romanes chose Lloyd Morgan, who put an end to all this unrestrained speculation. Morgan, a British psychologist, formulated in 1894 the probably most quoted recommendation in all of psychology:

> In no case may we interpret an action as the outcome of the exercise of a higher psychical faculty, if it can be interpreted as the outcome of the exercise of one which stands lower on the psychological scale.[16]

Generations of psychologists have dutifully repeated Morgan's Canon, taking it to mean that it is safe to assume that animals are stimulus-response machines. But Morgan never meant it that way. In fact, he rightly added, "But surely the simplicity of an explanation is no necessary criterion of its truth."[17] Here he was reacting against the mindset according to which animals are blind automata without souls. No self-respecting scientist would talk of "souls," but to deny animals *any* intelligence and consciousness came close enough. Taken aback by these views, Morgan added a provision to his canon according to which there is nothing wrong with more complex cognitive interpretations if the species in question has already been proven to have high intelligence.[18] With animals such as chimpanzees, elephants, and crows, for which we have ample evidence of complex cognition, we really do not need to start at zero every time we are struck by seemingly smart behavior. We don't need to explain their behavior the way we would that of, say, a rat. And even for the poor underestimated rat, zero is unlikely to be the best starting point.

Morgan's Canon was seen as a variation on Occam's razor, according to which science should seek explanations with the smallest number of assumptions. This is a noble goal indeed, but what if a minimalist cognitive explanation asks us to believe in miracles? Evolutionarily speaking, it would be a true miracle if we had the fancy cognition that we believe we have while our fellow animals had none of it. The pursuit of cognitive parsimony often conflicts with evolutionary parsimony.[19] No biologist is willing to go this far: we believe in gradual modification. We don't like to propose gaps between related species without at least coming up with an explanation. How did our species become rational and conscious if the rest of the natural world lacks any stepping-stones? Rigorously applied to animals—and to animals alone!—Morgan's Canon promotes a saltationist view that leaves the human mind dangling in empty evolutionary space. It is to the credit of Morgan himself that he recognized the limitations of his canon and urged us not to confuse simplicity with reality.

It is less known that ethology, too, arose amid skepticism about subjective methods. Tinbergen and other Dutch ethologists were shaped by the hugely popular illustrated books of two schoolmasters who taught love and respect for nature while insisting that the only way to truly understand animals was to watch them outdoors. This inspired a massive youth movement in Holland, with field excursions every Sunday, that laid the groundwork for a generation of eager naturalists. This approach did not combine well, however, with the Dutch tradition of "animal psychology," the dominant figure of which was Johan Bierens de Haan.

Internationally famous, erudite, and professorial, Bierens de Haan must have looked rather out of place as an occasional guest at Tinbergen's field site in the Hulshorst, a dune area in the middle of the country. While the younger generation ran around in shorts holding butterfly nets, the older professor came in suit and tie. These visits attest to the cordiality between both scientists before they grew apart, but young Tinbergen soon began to challenge the tenets of animal psychology, such as its reliance on introspection. Increasingly, he put distance between his own thinking and Bierens de Haan's subjectivism.[20] Not being from the same country, Lorenz showed less patience with the old man, whom he—in a play on his name—mischievously dubbed *Der Bierhahn* (German for "the beertap").

Tinbergen is nowadays best known for his Four Whys: four different yet complementary questions that we ask about behavior. But none of them explicitly mentions intelligence or cognition.[21] That ethology avoided any mention of internal states was perhaps essential for a budding empirical science. As a consequence, ethology temporarily closed the book on cognition and focused instead on the survival value of behavior. In doing so, it planted the seeds of sociobiology, evolutionary psychology, and behavioral ecology. This focus also offered a convenient way around cognition. As soon as questions about intelligence or emotions came up, ethologists would quickly rephrase them in functional terms. For example, if one bonobo reacted to the screams of another by rushing over for a tight embrace, classical ethologists will first of all wonder about the function of such behavior. They'd have debates about who benefited the most, the performer or the recipient, without asking what bonobos understood about one another's situations, or why the emotions of one should affect those of another. Might apes be empathic? Do bonobos evaluate one another's needs? This kind of cognitive query made (and still makes) many ethologists uncomfortable.

Blaming the Horse

It is curious that ethologists looked down on animal cognition and emotions as too speculative, while feeling on safe ground with behavioral evolution. If there is one area rife with conjecture, it is how behavior evolved. Ideally, you'd first establish the behavior's heredity and then measure its impact on survival and reproduction over multiple generations. But we rarely get anywhere close to having this information. With fast-breeding organisms, such as slime molds or fruit flies, these questions may be answerable, but evolutionary accounts of elephant behavior, or human behavior for that matter, remain largely hypothetical since these species don't permit large-scale breeding experiments. While we do have ways of testing hypotheses and mathematically modeling the consequences of behavior, the evidence is largely indirect. Birth control, technology, and medical care make our own species an almost hopeless test case for evolutionary ideas, which is why we have a plethora of speculations about what happened in the Environment of Evolutionary Adaptedness (EEA). This refers to the living conditions of our hunter-gatherer ancestors, about which we obviously have incomplete knowledge.

In contrast, cognition research deals with processes in real time. Even though we cannot actually "see" cognition, we are able to design experiments that help us deduce how it works while eliminating alternative accounts. In this regard, it really isn't different from any other scientific endeavor. Nevertheless, the study of animal cognition is still often considered a soft science, and until recently young scientists were advised away from such a tricky topic. "Wait until you have tenure," some older professors would say. The skepticism goes all the way back to the curious case of a German horse named Hans, who lived around the time Morgan crafted his canon. Hans became its proof positive. The black stallion was known in German as *Kluger Hans*, translated as Clever Hans, since he seemed to excel at addition and subtraction. His owner would ask

Clever Hans was a German horse that drew admiring crowds about a century ago. He seemed to excel at arithmetic, such as addition and multiplication. A more careful examination revealed, however, that his main talent was the reading of human body language. He succeeded only if he could see someone who knew the answer.

him to multiply four by three, and Hans would happily tap his hoof twelve times. He could also tell you what the date of a given weekday was if he knew the date of an earlier day, and he could tell the square root of sixteen by tapping four times. Hans solved problems he had never heard before. People were flabbergasted, and the stallion became an international sensation.

That is, until Oskar Pfungst, a German psychologist, investigated the horse's abilities. Pfungst had noticed that Hans was successful only if his owner knew the answer and was visible to the horse. If the owner or any other questioner stood behind a curtain while posing their question, the horse failed. It was a frustrating experiment for Hans, who would bite Pfungst if he got too many answers wrong. Apparently, the way he got them right is that the owner would subtly shift his position or straighten his back the moment Hans reached the correct number of taps. The questioner would be tense in face and posture until the horse reached the answer, at which point he would relax. Hans was very good at picking up these cues. The owner also wore a hat with a wide brim, which would be down as long as he looked at Hans's tapping hoof and go up when Hans reached the right number. Pfungst demonstrated that anyone wearing such a hat could get any number out of the horse by lowering and then raising his head.[22]

Some spoke of a hoax, but the owner was unaware that he was cuing his horse, so there was no fraud involved. Even once the owner knew, he found it nearly impossible to suppress his signals. In fact, following the report by Pfungst, the owner was so disappointed that he accused the horse of treachery and wanted him to spend the rest of his life pulling hearses as punishment. Instead of being mad at himself, he blamed his horse! Luckily for Hans, he ended up with a new owner who admired his abilities and tested them further. This was the right spirit, because instead of looking at the whole affair as a downgrading of animal intelligence, it proved incredible sensitivity. Hans's talent at arithmetic may have been flawed, but his understanding of human body language was outstanding.[23]

As an Orlov Trotter stallion, Hans appears to have perfectly fit the description of this Russian breed: "Possessed of amazing intelligence, they learn quickly and remember easily with few repetitions. There is often an uncanny understanding of what is wanted and needed of them at any given time. Bred to love people, they bond very tightly to their owners."[24]

Instead of being a disaster for animal cognition studies, the horse's exposé proved a blessing in disguise. Awareness of the Clever Hans Effect, as it became known, has greatly improved animal testing. By illustrating the power of blind procedures, Pfungst paved the way for cognitive studies that were able to withstand scrutiny. Ironically, this lesson is often ignored in research on humans. Young children are typically presented with cognitive tasks while sitting on their mothers' laps.

The assumption is that mothers are like furniture, but every mother wants her child to succeed, and nothing guarantees that her body movements, sighs, and nudges don't cue her child. Thanks to Clever Hans, the study of animal cognition is now more rigorous than that. Dog labs test the cognition of their animals while the human owner is blindfolded or stands in a corner while facing away. In one well-known study, in which Rico, a border collie, recognized more than two hundred words for different toys, the owner would ask for a specific toy located in a different room. This prevented the owner from looking at the toy and unconsciously guiding the dog's attention. Rico would need to run to the other room to fetch the mentioned item, which is how the Clever Hans Effect was avoided.[25]

We owe Pfungst a profound debt for demonstrating that humans and animals develop communication that they are unaware of. The horse reinforced behavior in his owner, and the owner in his horse, whereas everyone was convinced that they were doing something else entirely. While the realization of what was going on moved the historical pendulum to swing firmly from rich to lean interpretations of animal intelligence—where it unfortunately got stuck for too long—other appeals to simplicity have fared less well. Below I describe two examples, one concerning *self-awareness* and the other *culture*, both concepts that, whenever mentioned in relation to animals, still send some scholars through the roof.

Armchair Primatology

When American psychologist Gordon Gallup, in 1970, first showed that chimpanzees recognize their own reflection, he spoke of self-awareness—a capacity that he said was lacking in species, such as monkeys, that failed his mirror test.[26] The test consisted of putting a mark on the body of an anesthetized ape that it could find only, once awake, by inspecting its reflection. Gallup's choice of words obviously annoyed those leaning toward a robotic view of animals.

The first counterattack came from B. F. Skinner and colleagues, who promptly trained pigeons to peck at dots on themselves while standing in front of a mirror.[27] Reproducing a semblance of the behavior, they felt, would solve the mystery. Never mind that it took them hundreds of grain rewards to get the pigeons to do something that chimpanzees and humans do without any coaching. One can train goldfish to play soccer and bears to dance, but does anyone believe that this tells us much about the skills of human soccer stars or dancers? Worse, we aren't even sure that this pigeon study is replicable. Another research team spent years trying the exact same training, using the same strain of pigeon, without producing any self-pecking birds. They ended up publishing a report critical of the original study with the word *Pinocchio* in its title.[28]

The second counterattack was a fresh interpretation of the mirror test, suggesting that the observed self-recognition might be a by-product of the anesthesia used in the marking procedure. Perhaps when a chimpanzee recovers from the anesthesia, he randomly touches his face, resulting in accidental contact with the mark.[29] This idea was quickly disproved by another team that carefully recorded which facial areas chimpanzees touch. It turned out that the touching is far from random: it specifically targets the marked area and peaks right after the ape has seen his own reflection.[30] This was, of course, what the experts had been saying all along, but now it was official.

Apes really don't need anesthesia to show how well they understand mirrors. They spontaneously use them to look inside their mouth, and females always turn around to check out their behinds—something males don't care about. Both are body parts that they normally never get to see. Apes also use mirrors for special needs. For example, Rowena has a little injury on the top of her head caused by a scuffle with a male. Immediately, when we hold up a mirror, she inspects the injury and

B. F. Skinner was more interested in experimental control over animals than spontaneous behavior. Stimulus-response contingencies were all that mattered. His behaviorism dominated animal studies for much of the last century. Loosening its theoretical grip was a prerequisite for the rise of evolutionary cognition.

grooms around it while following the reflection of her movements. Another female, Borie, has an ear infection that we are trying to treat with antibiotics, but she keeps waving her hand in the direction of a table that is empty except for a small plastic mirror. It takes a while before we understand her intentions, but as soon as we hand her the toy, she picks up a straw and angles the mirror such that she can clean out her ear while watching the process in the mirror.

A good experiment doesn't create new and unusual behavior but taps into natural tendencies, which is exactly what Gallup's test did. Given the apes' spontaneous mirror use, no expert would ever have come up with the anesthesia story. So what makes scientists unaccustomed to primates think they know better? Those of us who work with exceptionally gifted animals are used to unsolicited opinions about how we ought to test them and what their behavior actually means. I find the arrogance behind such advice mind-boggling. Once, in his desire to underscore the uniqueness of human altruism, a prominent child psychologist shouted at a large audience, "No ape will ever jump into a lake to save another!" It was left to me to point out during the Q&A afterward that there are actually a handful of reports of apes doing precisely this—often to their own detriment, since they don't swim.[31]

The same arrogance explains the doubts raised about one of the best-known discoveries in field primatology. In 1952 the father of Japanese primatology, Kinji Imanishi, first proposed that we may justifiably speak of animal culture if individuals learn habits from one another resulting in behavioral diversity between groups.[32] By now fairly well accepted, this idea was so radical at the time that it took Western science forty years to catch up. In the meantime, Imanishi's students patiently documented the spreading of sweet potato washing by Japanese macaques on Koshima Island. The first monkey to do so was a juvenile female, named Imo, now honored with a statue at the entrance to the island. From Imo the habit spread to her age peers, then to their mothers, and eventually to nearly all monkeys on the island. Sweet potato washing became the best-known example of a learned social tradition, passed on from generation to generation.

Many years later, this view triggered a so-called *killjoy account*—an attempt to deflate a cognitive claim by proposing a seemingly simpler alternative—according to which the monkey-see-monkey-do explanation of Imanishi's students was overblown. Why couldn't it just have been individual learning—that is, each monkey acquired potato washing on its own without the assistance of anybody else? There might even have been human influence. Perhaps potatoes were handed out selectively by Satsue Mito, Imanishi's assistant, who knew every monkey by name. She may have rewarded monkeys who dipped their spuds in the water, thus prompting them to do so ever more frequently.[33]

The only way to find out was to go to Koshima and ask. Having been twice to this island in the subtropical south of Japan, I had a chance to interview the then eighty-four-year-old Mrs. Mito via an interpreter.

The first evidence for animal culture came from sweet-potato-washing Japanese macaques on Koshima Island. Initially, the washing tradition spread among same-aged peers, but nowadays it is propagated transgenerationally, from mother to offspring.

She reacted with incredulity to my question about food provisioning. One cannot hand out food any way one wants, she insisted. Any monkey that holds food while high-ranking males are empty-handed risks getting into trouble. Macaques are very hierarchical and can be violent, so putting Imo and other juveniles before the rest would have endangered their lives. In fact, the last monkeys to learn potato washing, the adult males, were the first ones to be fed. When I brought up the argument to Mrs. Mito that she might have rewarded washing behavior, she denied that this was even possible. In the early years, potatoes were handed out in the forest far away from the freshwater stream where the monkeys did their cleaning. They'd collect their spuds and quickly run off with them, often bipedally since their hands were full. There was no way for Mito to reward whatever they did in the distant stream.[34] But perhaps the strongest argument for social as opposed to individual learning was the way the habit spread. It can hardly be coincidental that one of the first to follow Imo's example was her mother, Eba. After this, the habit spread to Imo's peers. The learning of potato washing nicely tracked the network of social relations and kinship ties.[35]

Like the scientist who gave us the mirror-anesthesia hypothesis, the one who wrote an entire article debunking the Koshima discovery was a nonprimatologist who, moreover, never bothered to set foot on Koshima or check his ideas with the fieldworkers who had camped for decades on the island. Again, I can't help but wonder about the mismatch between conviction and expertise. Perhaps this attitude is a leftover of the mistaken belief that if you know enough about rats and pigeons, you know everything there is to know about animal cognition. It prompts me to propose the following know-thy-animal rule: *Anyone who wishes to stress an alternative claim about an animal's cognitive capacities either needs to familiarize him- or herself with the species in question or make a genuine effort to back his or her counterclaim with data.* Thus, while I admire Pfungst's work with Clever Hans and its eye-opening conclusions, I have great trouble with armchair speculations devoid of any attempt to check their validity. Given how seriously the field of evolutionary cognition takes variation between species, it is time to respect the special expertise of those who have devoted a lifetime getting to know one of them.

The Thaw

One morning at Burgers' Zoo, we showed the chimpanzees a crate full of grapefruits. The colony was in the building where it spends the night, which adjoins a large island, where it spends the day. The apes seemed interested enough watching us carry the crate through a door onto the island. When we returned to the building with an empty crate, however, pandemonium broke out. As soon as they saw that the fruits were gone, twenty-five apes burst out hooting and hollering in a most festive mood, slapping one another's backs. I have never seen animals so excited about *absent* food. They must have inferred that grapefruits cannot vanish, hence must have remained on the island onto which the colony would soon be released. This kind of reasoning does not fall into any simple category of trial-and-error learning, especially since it was the first time we followed this procedure. The grapefruit experiment was a one-time event to study responses to cached food.

One of the first tests of *inferential reasoning* was conducted by American psychologists David and Ann Premack, who presented Sadie, a chimpanzee, with two boxes. They placed an apple in one and a banana in the other. After a few minutes of distraction, Sadie saw one of the experimenters munching on either an apple or a banana. This experimenter then left, and Sadie was released to inspect the boxes. She faced an interesting dilemma, since she had not seen how the experimenter had gotten his fruit. Invariably, Sadie would go to the box with the fruit that the experimenter had *not* eaten. The Premacks ruled out gradual learning, because Sadie made this choice on the very first trial as well as all subsequent ones. She seemed to have reached two conclusions. First, that the eating experimenter had removed his fruit from one of the two boxes, even if she had not actually seen him do so. And second, that this meant that the other box must still hold the other fruit. The Premacks note that most animals don't make any such assumptions: they just see an experimenter consume fruit, that's all. Chimpanzees, in contrast, try to figure out the order of events, looking for logic, filling in the blanks.[36]

Years later the Spanish primatologist Josep Call presented apes with two covered cups. They had learned that only one would be baited with grapes. If Call removed the tops and let them look inside the cups, the apes chose the one with grapes. Next, he kept the cups covered and shook first one, then the other. Only the cup with grapes made noise, which was the one they preferred. This was not too surprising. But making things harder, Call would sometimes shake only the empty cup, which made no noise. In this case, the apes would still pick the other one, thus operating on the basis of exclusion. From the absence of sound, they guessed where the grapes must be. Perhaps we are not impressed by this either, as we take such inferences for granted, but it is not all that obvious. Dogs, for example, flunk this task. Apes are special in that they seek logical connections based on how they believe the world works.[37]

Here it gets interesting, because aren't we supposed to go for the simplest possible explanation? If large-brained animals, such as apes, try to get at the logic behind events, could this be the simplest level at which they operate?[38] It reminds me of Morgan's provision to his canon, according to which we are allowed more complex premises in the case of more intelligent species. We most certainly apply this rule to ourselves. We always try to figure things out, applying our reasoning powers to everything around us. We go so far as to invent causes if we can't find any, leading to weird superstitions and supernatural beliefs, such as sports fans wearing the same T-shirt over and over for luck, and disasters being blamed on the hand of God. We are so logic-driven that we can't stand the absence of it.

Evidently, the word *simple* is not as simple as it sounds. It means different things in relation to different species, which complicates the eternal battle between skeptics and cognitivists. In addition, we often get tangled up in semantics that aren't worth the heat they generate. One scientist will argue that monkeys understand the danger posed by leopards, whereas another will say that monkeys have merely learned from experience that leopards sometimes kill members of their species. Both statements are really not that different, even though the first uses the language of understanding, and the second of learning. With the decline of behaviorism, debates on these issues have fortunately grown less fiery. By attributing all behavior under the sun to a single learning mechanism, behaviorism set up its own downfall. Its dogmatic overreach made it more like a religion than a scientific approach. Ethologists loved to slam it, saying that instead of domesticating white rats in order to make them suitable to a particular testing paradigm, behaviorists should have done the opposite. They should have invented paradigms that fit "real" animals.

The counterpunch came in 1953, when Daniel Lehrman, an American comparative psychologist, sharply attacked ethology.[39] Lehrman objected to simplistic definitions of *innate*, saying that even species-typical behavior develops from a history of interaction with the environment. Since nothing is purely inborn, the term *instinct* is in fact misleading and should be avoided. Ethologists were stung and dismayed by his unexpected critique, but once they got over their "adrenaline attack" (Tinbergen's words), they discovered that Lehrman hardly fit the behaviorist bogeyman stereotype. He was an enthusiastic bird-watcher, for example, who knew his animals. This impressed the ethologists, and Baerends recalled that while meeting the "enemy" in person, they managed to resolve most misunderstandings, found common ground, and became "very good friends."[40] Once Tinbergen became acquainted with Danny, as they now knew Lehrman, he went so far as to call him more of a zoologist than a psychologist, which the latter took as a compliment.[41]

Their bonding over birds went way beyond the way John F. Kennedy and Nikita Khrushchev bonded over Pushinka, a little dog that the Soviet leader sent to the White House. Despite this gesture, the Cold War continued unabated. In contrast, Lehrman's harsh critique and the subsequent meeting of minds between comparative psychologists and ethologists set in motion a process of mutual respect and understanding. Tinbergen, in particular, acknowledged Lehrman's influence on his later thinking. Apparently, they had needed a big spat to start a rapprochement, which was hastened by ongoing criticism *within* each camp of its *own* tenets. Within ethology, the younger generation grumbled about the rigid Lorenzian drive and instinct concepts, whereas comparative psychology had an even longer tradition of challenges to its own dominant paradigm.[42] Cognitive approaches had been tried off and on, even as early as the 1930s.[43] But ironically, the biggest blow to behaviorism came from within. It all started with a simple learning experiment conducted on rats.

Anyone who has tried to punish a dog or cat for problematic behavior knows that it is best to do so quickly, while the offense is still visible or

American psychologist Frank Beach lamented the narrow focus of behavioral science on the albino rat. His incisive critique featured a cartoon in which a Pied Piper rat is followed by a happy mass of white-coated experimental psychologists. Carrying their favorite tools—mazes and Skinner boxes—they are being led into a deep river. After S. J. Tatz in Beach (1950).

at least fresh in the animal's mind. If you wait too long, your pet doesn't connect your scolding with the stolen meat or the droppings behind the couch. Since short intervals between behavior and consequence have always been regarded as essential, no one was prepared when, in 1955, the American psychologist John Garcia claimed he had found a case that broke all the rules: rats learn to refuse poisoned foods after just a single bad experience even if the resulting nausea takes hours to set in.[44] Moreover, the negative outcome had to be nausea—electric shock didn't have the same effect. Since toxic nutrition works slowly and makes you sick, none of this was particularly surprising from a biological standpoint. Avoiding bad food seems a highly adaptive mechanism. For standard learning theory, however, these findings came like a bolt out of the blue, due to the assumption that time intervals should be short while the kind of punishment is irrelevant. The findings were in fact devastating, and

Garcia's conclusions were so unwelcome that he had trouble getting them published. One imaginative reviewer contended that his data were less likely than finding bird shit in a cuckoo clock! The *Garcia Effect* is now well established, though. In our own lives, we remember food that has poisoned us so well that we gag at the mere thought of it or never set foot in a certain restaurant again.

For readers who wonder about the fierce resistance to Garcia's discovery despite the fact that most of us have firsthand experience with the power of nausea, it is good to realize that human behavior was (and still is) often seen as the product of reflection, such as an analysis of cause and effect, whereas animal behavior was supposedly free of such processes. Scientists were not ready to equate the two. Human reflection is chronically overrated, though, and we now suspect that our own reaction to food poisoning is in fact similar to that of rats. Garcia's findings forced comparative psychology to admit that evolution pushes cognition around, adapting it to the organism's needs. This became known as *biologically prepared learning*: each organism is driven to learn those things it needs to know in order to survive. This realization obviously helped the rapprochement with ethology. Moreover, the geographic distance between both schools fell away. Once comparative psychology took hold in Europe—which is how I briefly ended up in a behaviorist lab—and ethology was being taught in North American zoology departments, students on both sides of the Atlantic could absorb the entire range of views and begin to integrate them. The synthesis between the two approaches did not take place just at international meetings or in the literature, therefore, but also in the classrooms.

We entered a period of crossover scholars, which I'll illustrate with just two examples. The first is the American psychologist Sara Shettleworth, who for most of her career taught at the University of Toronto, and who has been influential through her textbooks on animal cognition. She started out in the behaviorist corner but ended up advocating a biological approach to cognition that is sensitive to the ecological needs of each species. She remains as cautious in her interpretations of cognition as one would expect from someone of her background, yet her work gained a clear ethological flavor, which she attributes to certain professors when she was a student as well as involvement with her husband's fieldwork on sea turtles. In an interview about her career, Shettleworth explicitly mentions Garcia's work as a turning point that opened the eyes of her field to the evolutionary forces shaping learning and cognition.[45]

At the other end of the scale is one of my heroes, Hans Kummer, a Swiss primatologist and ethologist. As a student, I avidly devoured every paper he wrote, mostly his field studies on hamadryas baboons in Ethiopia. Kummer did not just observe social behavior and relate it to ecology; he always puzzled about the cognition behind it and conducted field experiments on (temporarily) captured baboons. He later moved to captive work on long-tailed macaques at the University of Zürich. Kummer felt that the only way to test cognitive theories was by means of controlled experiments. Observation alone was not going to cut it, so primatologists should become more like comparative psychologists if they ever wished to unravel the puzzle of cognition.[46]

I went through a similar transition from observation to experimentation and was greatly inspired by Kummer's macaque lab when I set up my own lab for capuchin monkeys. The trick is to house the animals socially, hence build large indoor and outdoor areas, where the monkeys can spend most of the day playing, grooming, fighting, catching insects, and so on. We trained them to enter a test chamber where they could work on a touchscreen or a social task before we'd return them to the group. This arrangement had two advantages over traditional labs, which keep monkeys, rather like Skinner's pigeons, in single cages. First of all, there is the quality of life issue. It is my personal feeling that if we are going to keep highly social animals in captivity, the very least we can do for them is permit them a group life. This is the best and most ethical way to enrich their lives and make them thrive.

Second, it makes no sense to test monkeys on social skills without giving them a chance to express these skills in daily life. They need to be completely familiar with one another for us to investigate how they share food, cooperate, or judge one another's situation. Kummer understood all this, having started out, like myself, as a primate watcher. In my opinion, anyone who intends to conduct experiments on animal cognition should first spend a couple thousand hours observing the spontaneous behavior of the species in question. Otherwise we get experiments uninformed by natural behavior, which is precisely the approach we should be leaving behind.

Today's evolutionary cognition is a blend of both schools, taking the best parts of each. It applies the controlled experimental methodology developed by comparative psychology combined with the blind testing that worked so well with Clever Hans, while adopting the rich evolutionary framework and observation techniques of ethology. For many young scientists, it is now immaterial whether we call them comparative psychologists or ethologists, since they integrate concepts and techniques from both. On top of this comes a third major influence, at least for work in the field. The impact of Japanese primatology is not always recognized in the West—which is why I have called it a "silent invasion"—but we routinely name individual animals and track their social careers across multiple generations. This allows us to understand the kinship ties and friendships at the core of group life. Begun by Imanishi right after World War II, this method has become standard in work on long-lived mammals, from dolphins to elephants and primates.

Unbelievably, there was a time when Western professors warned their students away from the Japanese school because naming animals was considered too humanizing. There was of course also the language barrier, which made it hard for Japanese scientists to get heard. Junichiro Itani, Imanishi's foremost student, was met with disbelief when he toured American universities in 1958 because no one believed that he and his colleagues were able to tell a hundred or more monkeys apart. Monkeys look so much alike that Itani obviously was making things up. He once told me that he was mocked to his face and had no one to defend him except the great American primatological pioneer Ray Carpenter, who did see the value of this approach.[47] Nowadays, of course, we know that recognizing a large number of monkeys is possible, and we all do it. Not unlike Lorenz's emphasis on knowing the whole animal, Imanishi urged us to empathize with the species under study. We need to get under its skin, he said, or as we would nowadays put it, try to enter its *Umwelt*. This old theme in the study of animal behavior is quite different from the misguided notion of critical distance, which has given us excessive worries about anthropomorphism.

The eventual international embrace of the Japanese approach illustrates something else that we learned from the tale of two schools— ethology and comparative psychology—which is that the initial animosity between divergent approaches can be overcome if we realize that each has something to offer that the other lacks. We may weave them together into a new whole that is stronger than the sum of its parts. The fusing of complementary strands is what makes evolutionary cognition the promising approach it is today. But sadly it took a century of misunderstandings and colliding egos before we got there.

Beewolves

Tinbergen was in tears when I last saw him. It was 1973, the year in which he, Lorenz, and von Frisch were honored with Nobel prizes. He had come to Amsterdam to receive a different medal and give a lecture. Speaking in Dutch, his voice quavering with emotion, he asked what we had done to his country. The magnificent little spot in the dunes where he had studied gulls and terns was no more. Decades earlier, while emigrating aboard a boat to England, he had pointed at the site—the eternal self-rolled cigarette in his hand—predicting that "it will all go, irrevocably." Years later the place was swallowed up by the expansion of Rotterdam harbor, then the busiest in the world.[48]

Tinbergen's lecture reminded me of all the great things he had done, which included animal cognition, even though he never used the term. He had worked on how digger wasps find their nest after a trip away. Also known as beewolves, these wasps capture and paralyze a honeybee, drag it to their nest in the sand (a long burrow), and leave it as a meal for their larvae. Before they go out to hunt for a bee, they make a brief orientation flight to memorize the location of their inconspicuous burrow. Tinbergen put objects around the nest, such as a circle of pinecones, to see what information they used to find it back. He was able to trick the wasps, making them search at the wrong location, by moving his pinecones around.[49] His study addressed problem solving tied to a species' natural history, precisely the topic of evolutionary cognition. The wasps proved very good at this particular task.

Brainier animals have less restricted cognition and often find solutions to novel or unusual problems. The ending of my grapefruit story with the chimpanzees offers a nice demonstration. After releasing the apes onto the island, a number of them passed over the site where we had hidden the fruits under the sand. Only a few small yellow patches were visible. Dandy, a young adult male, hardly slowed down when he ran over the place. Later in the afternoon, however, when all the apes were dozing off in the sun, he made a beeline for the spot. Without hesitation, he dug up the fruits and devoured them at his leisure, which he would never have been able to do had he stopped right when he saw them. He would have lost them to dominant group mates.[50]

Here we see the entire spectrum of animal cognition, from the specialized navigation of a predatory wasp to the generalized cognition of apes, which allows them to handle a great variety of problems, including novel ones. What struck me most is that Dandy at his first passing didn't linger for a second. He must have made an instant calculation that deception was going to be his best bet.

3 | COGNITIVE RIPPLES

Eureka!

The sunny, breezy Canary Islands are about the last place in the world where one would expect a cognitive revolution, yet this is where it all began. In 1913 the German psychologist Wolfgang Köhler came to Tenerife, off the coast of Africa, to head the Anthropoid Research Station, where he remained until after World War I. Even though rumor has it that his job was to spy on passing military vessels, Köhler devoted most of his attention to a small colony of chimpanzees.

Having eluded indoctrination in the learning theories of his day, Köhler was refreshingly open-minded about animal cognition. Instead of trying to control his animals to seek specific outcomes, he had a wait-and-see attitude. He presented them with simple challenges to find out how they'd meet them. For his most talented chimpanzee, Sultan, he would put a banana out of reach on the ground and offer him sticks that were too short to reach the fruit. Or he would hang a banana high up in the air and spread large wooden boxes around, none of which was tall enough for the purpose. Sultan would first jump or throw things at the banana or drag humans by the hand toward it in the hope that they'd help him out, or at least be willing to serve as a footstool. If this failed,

Grande, a female chimpanzee, piles up four boxes to reach a banana. A century ago Wolfgang Köhler set the stage for animal cognition research by demonstrating that apes can solve problems in their heads by means of a flash of insight, before enacting the solution.

he would sit around for a while without doing anything until he might hit at a sudden solution. He would jump up to put one bamboo stick inside another, making a longer stick. Or he would stack boxes on top of one another so as to build a tower that allowed him to reach the banana. Köhler described this moment as the "aha! experience," as if a lightbulb had been switched on, not unlike the story of Archimedes, who jumped out of his bath in which he had discovered how to measure the volume of submerged objects, after which he ran naked through the streets of Syracuse, shouting "Eureka!"

According to Köhler, a sudden *insight* explained how Sultan put together what he knew about bananas, boxes, and sticks to produce a brand-new action sequence that would take care of his problem. The scientist ruled out imitation and trial-and-error learning, since Sultan had had no previous experience with these solutions nor ever been rewarded for them. The outcome was "unwaveringly purposeful" action in which the ape kept trying to reach his goal despite the numerous stacking errors resulting in the collapse of his towers. A female, Grande, was an even more undeterred and patient architect who once built a wobbly tower of four boxes. Köhler remarked that once a solution was discovered, the apes found it easier to solve similar problems, as if they had learned something about the causal connections. He described his experiments in admirable detail in *The Mentality of Apes* in 1925, which was at first ignored and then disparaged, but that now stands as a classic in evolutionary cognition.[1]

The insightful solutions of Sultan and other apes hint at the kind of mental activity that we refer to as "thinking," even though its precise nature was (and still is) barely understood. A few years later the American primate expert Robert Yerkes described similar solutions.

> *Frequently I have seen a young chimpanzee, after trying in vain to get its reward by one method, sit down and reexamine the situation as though taking stock of its former efforts and trying to decide what to do next. . . . More startling by far than the quick passage from one method to another, the definiteness of acts, or the pauses between efforts, is the sudden solution of problems. . . . Frequently, although not in all individuals or in all problems, correct and adequate solution is achieved without warning and almost instantly.[2]*

Yerkes went on to note that those who only know animals that are good at trial-and-error learning "can scarcely be expected to believe" his descriptions. He thus anticipated the inevitable pushback to these revolutionary ideas. Unsurprisingly, it arrived in the form of trained pigeons shoving little boxes around in a dollhouse so that they could stand on top of them to reach a tiny plastic banana associated with grain rewards.[3] How entertaining! At the same time, Köhler's interpretations were criticized as anthropomorphic. But I heard an amusing antidote to these accusations from an American primatologist brave enough to enter the Skinnerian lion's den in the 1970s, where he debated tool-using apes.

Without offering specifics, Emil Menzel told me that an eminent East Coast professor once invited him to speak. This professor looked down on primate research and was openly hostile to cognitive interpretations, two orientations that often go together. Perhaps he invited young Menzel to make fun of him, not realizing that the tables might be turned. Menzel treated his audience to spectacular footage of his chimpanzees putting a long pole against their enclosure's high wall. While some individuals held the pole steady, others scaled it to reach temporary freedom. It was a complex operation since the apes needed to avoid coils of electrified wire while recruiting one another's assistance at critical moments through hand gestures. Menzel, who had filmed all this himself, decided to run his footage without mentioning intelligence. He was going to be as neutral as possible. His narration was purely descriptive: "You now see Rock grab the pole while glancing at the others," or "Here a chimpanzee swings over the wall."[4]

After his lecture, the professor jumped up to accuse Menzel of being unscientific and anthropomorphic, of attributing plans and intentions to animals that obviously had neither. To a roar of approval, Menzel countered that he had not attributed anything. If this professor had seen plans and intentions, he must have seen them with his own eyes, because Menzel himself had refrained from suggesting any such things.

Interviewing Menzel at my home (he lived nearby) a few years before his death, I took the opportunity to ask him about Köhler. Widely recognized as a great expert on great apes himself, Menzel said it had taken him years working with chimps to fully appreciate this pioneer's genius. Like Köhler, Menzel believed in watching over and over and thinking through what his observations might mean, even if he'd seen a certain behavior only once. He protested against labeling a single observation an "anecdote," adding with a mischievous smile, "My definition of an anecdote is someone else's observation." If you have seen something yourself and followed the entire dynamic, there usually is no doubt in your mind of what to make of it. But others may be skeptical and need convincing.

Here I cannot resist telling an anecdote of my own. And I do not mean The Great Escape at Burgers' Zoo, where the chimpanzee colony did exactly what Menzel had documented. After twenty-five apes raided the zoo's restaurant, we found a tree trunk, far too heavy for a single ape to carry, propped against the inside wall of their enclosure. No, I mean an insightful solution to a *social* problem—a sort of social tool use— that is my specialty. Two female chimps were sitting in the sun, with their children rolling around in the sand in front of them. When the play turned into a screaming, hair-pulling fight, neither mother knew what to do because if one of them tried to break up the fight, it was guaranteed that the other would protect her offspring, since mothers are never impartial. It is not unusual for a juvenile quarrel to escalate into an adult fight. Both mothers nervously monitored each other as well as the fight. Noticing the alpha female, Mama, asleep nearby, one of them went over to poke her in the ribs. As the old matriarch got up, the mother pointed at the fight by swinging an arm in its direction. Mama needed only one glance to grasp what was going on and took a step forward with a threatening grunt. Her authority was such that this shut up the youngsters. The mother had found a quick and efficient solution to her problem, relying on the mutual understanding typical of chimpanzees.

Similar understanding can be seen in their altruism, such as when younger females collect water in their mouths for an aging female, who can barely walk anymore, spitting it into her open mouth so that she doesn't have to walk all the way to the spigot. The British primatologist Jane Goodall described how Madame Bee, a wild chimpanzee, had become too old and weak to climb into fruiting trees. She would patiently wait at the bottom for her daughter to carry down fruits, upon which the two of them would contentedly munch together.[5] In such cases, too, apes grasp a problem and come up with a fresh solution, but the striking part here is that they perceive *another* ape's problem. Since these social perceptions have attracted much research, we'll delve into them later on, but let me clarify one general point about problem solving. Although Köhler stressed that trial-and-error learning could not explain his observations, it was not as if learning played no role at all. In fact, his apes committed tons of "stupidities," as Köhler called them, that showed that solutions were rarely perfectly formed in their minds and required quite a bit of tweaking.

His apes had undoubtedly learned the *affordances* of various items. This term from cognitive psychology refers to how objects can be used, such as the handle on a teacup (which affords holding) or the steps on a ladder (which afford climbing). Sultan must have known the affordances of sticks and boxes before he hit on his solutions. Similarly, the female chimp who activated Mama had no doubt witnessed the latter's effectiveness as arbitrator. Insightful solutions invariably rest on prior information. What is special about apes is their capacity to flexibly weave such preexisting knowledge into new patterns, never tried before, that work to their advantage. I have speculated the same about their political strategies, such as the way chimps will isolate a rival from his supporters or encourage a truce by dragging reluctant former combatants toward each other.[6] In all such cases, we see apes finding insightful solutions to everyday problems. They are so good at it that even the staunchest skeptic, as Menzel discovered, finds it impossible to watch them without being struck by their obvious intentionality and intelligence.

Wasp Mugs

There was a time when scientists thought behavior derived from either learning or biology. Human behavior was on the learning side, animal behavior on the biology side, and there was little in between. Never mind the false dichotomy (in all species, behavior is a product of both), but increasingly a third explanation had to be added: cognition. Cognition relates to the kind of information an organism gathers and how it processes and applies this information. Clark's nutcrackers remember where they have stored thousands of nuts, beewolves make an orientation flight before leaving their burrow, and chimpanzees nonchalantly learn the affordances of play objects. Without any reward or punishment, animals accumulate knowledge that will come in handy in the future, from finding nuts in the spring, to returning to one's burrow, to reaching a banana. The role of learning is obvious, but what is special about cognition is that it puts learning in its proper place. Learning is a mere tool. It allows animals to collect information in a world that, like the Internet, contains a staggering amount of it. It is easy to drown in the information swamp. An organism's cognition narrows down the information flow and makes it learn those specific contingencies that it needs to know given its natural history.

Many animals have cognitive achievements in common. The more scientists discover, the more ripple effects we notice. Capacities that were once thought to be uniquely human, or at least uniquely Hominoid (the tiny primate family of humans plus apes), often turn out to be widespread. Traditionally, apes have been the first to inspire discoveries thanks to their manifest intellect. After the apes break down the dam between humans and the rest of the animal kingdom, the floodgates often open to include species after species. Cognitive ripples spread from apes to monkeys to dolphins, elephants, and dogs, followed by birds, reptiles, fish, and sometimes invertebrates. This historical progression is not to be confused with a scale with Hominoids on top. I rather view it

Paperwasps live in small hierarchical colonies in which it pays to recognize every individual. Their black-and-yellow facial markings allow them to tell one another apart. A closely related wasp species with a less differentiated social life lacks face recognition, which shows how much cognition depends on ecology.

as an ever-expanding pool of possibilities in which the cognition of, say, the octopus may be no less astonishing than that of any given mammal or bird.

Consider face recognition, which was initially viewed as uniquely human. Now apes and monkeys have joined the countenance elite. Every year when I visit Burgers' Zoo, in Arnhem, a few chimps still remember me from more than three decades before. They pick out my face from the crowd, greeting me with excited hooting. Not only do primates recognize faces, but faces are special to them. Like humans, they show an "inversion effect": they have trouble recognizing faces that are turned upside down. This effect is specific for faces; how an image is oriented hardly matters for the recognition of other objects, such as plants, birds, or houses.

When we tested capuchin monkeys using touchscreens, we noticed that they freely tapped all sorts of images, but they freaked out at the first face that appeared. They clutched themselves and whined, reluctant to touch the picture. Did they treat it with more respect because putting a hand on a face violates a social taboo? Once they got over their hesitation, we showed them portraits of group mates and unknown monkeys.

All these portraits look alike to naïve humans since they concern the same species, but our monkeys had no trouble telling them apart, indicating with a little tap on the screen which ones they knew and which ones they didn't.[7] We humans take this ability for granted, but the monkeys had to link a two-dimensional pattern of pixels to a live individual in the real world, which they did. Face recognition, science concluded, is a specialized cognitive skill of primates. But no sooner had it done so than the first cognitive ripples arrived. Face recognition has been found in crows, sheep, even wasps.

It is unclear what faces mean to crows. In their natural lives, they have so many other ways of recognizing one another by calls, flight patterns, size, and so on, that faces may not be relevant. But crows have incredibly sharp eyes, so they likely notice that humans are easiest recognized by their faces. Lorenz reported harassment of certain people by crows and was so convinced of their ability to hold a grudge that he disguised himself with a costume whenever he captured and banded his jackdaws. (Both jackdaws and crows are corvids, a brainy bird family that also includes jays, magpies, and ravens.) Wildlife biologist John Marzluff at the University of Washington, in Seattle, has captured so many crows that these birds take his name in vain whenever he walks around, scolding and dive-bombing him, doing justice to the "murder" label used for a whole bunch of them.

> We don't know how they pick us out of the forty thousand folk scurrying like two-legged ants over well-worn trails. But single us out they do, and nearby crows flee while uttering a call that sounds to us like vocal disgust. In contrast, they calmly walk among our students and colleagues who have never captured, measured, banded, or otherwise humiliated them.[8]

Marzluff set out to test this recognition with rubber face masks like those we put on at Halloween. After all, crows may recognize certain people by their bodies, hair, or clothes, but with masks you can move a human "face" around from one body to the next, isolating its specific role. His angry birds experiment involved capturing crows while wearing a particular mask, then have coworkers walk around with either this mask or a neutral one. The crows easily remembered the mask of the capturer, far from fondly. Funny enough, the neutral mask was Vice President Dick Cheney's face, which elicited more negative reactions from the students on campus than from the crows. Not only did birds that had never been captured recognize the "predator" mask, but years later they still harassed its wearers. They must have picked up on the hateful response of their fellows resulting in massive distrust of specific humans. As Marzluff explains, "It would be a rare hawk that would be nice to a crow, but with humans you have to classify us as individuals. Clearly, they're able to do that."[9]

While corvids always impress us, sheep seem to go a step further in that they remember one another's faces. British scientists led by Keith Kendrick taught sheep the difference between twenty-five pairs of their own species' faces by rewarding a choice for one face and not for the other. To us, all these faces look eerily alike, but the sheep learned and retained the twenty-five differences for up to two years. In doing so, they used the same brain regions and neural circuits as humans, with some neurons responding specifically to faces and not to other stimuli. These special neurons were activated if the sheep saw pictures of companions that they remembered—they actually called out to these pictures as if the individuals were present. Publishing their study under the subtitle "sheep are not so stupid after all"—a title to which I object, since I don't believe in stupid animals—the investigators put the face-recognition ability of sheep on a par with that of primates and speculated that a flock, which to us looks like an anonymous mass, is in fact quite differentiated. This also means that mixing flocks, as is sometimes done, may cause more distress than we realize.

Having made primate chauvinists look sheepish, science piled it on with wasps. The northern paperwasp, common in the American Midwest, has a highly structured society with a hierarchy among its founding queens, who are dominant over all workers. Given the intense competition, each wasp needs to know her place. The alpha queen lays most eggs, followed by the beta queen, and so on. Members of the small colony are aggressive to outsiders as well as to females whose facial markings have been altered by experimenters. They recognize one another by strikingly different patterns of yellow and black on every female's face. The American scientists Michael Sheehan and Elizabeth Tibbetts tested individual recognition and found it to be as specialized as that of primates and sheep. The wasps distinguish their own species' mugs far better than other visual stimuli, and they also outperform a closely related wasp that lives in colonies founded by a single queen. These wasps hardly have a hierarchy and have far more homogeneous faces. They don't need individual recognition.[10]

If face recognition has evolved in such disparate pockets of the animal kingdom, one wonders how these capacities connect. Wasps do not have the big brains of primates and sheep—they have minuscule sets of neural ganglia—hence they must be doing it in a different manner. Biologists never tire of stressing the distinction between *mechanism* and *function*: it is very common for animals to achieve the same end (function) by different means (mechanism). Yet with respect to cognition, this distinction is sometimes forgotten when the mental achievements of large-brained animals are questioned by pointing at "lower" animals doing something similar. Skeptics delight in asking "If wasps can do it, what's the big deal?" This race to the bottom has given us trained pigeons hopping onto little boxes to disparage Köhler's experiments on apes and the holding up of intelligence outside the primate order to cast doubt on mental continuity between humans and other Hominoids.[11] The underlying thought is that of a linear cognitive scale, and the argument that since we rarely assume complex cognition in "lower" animals, there is no reason to do so in "higher" ones.[12] As if there were only one way to achieve a given outcome!

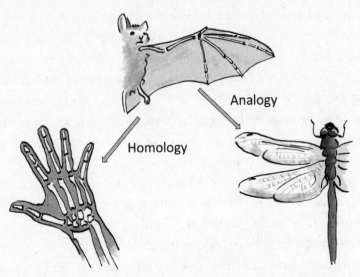

Evolutionary science distinguishes between homology (the traits of two species derive from a common ancestor) and analogy (similar traits evolved independently in two species). The human hand is homologous with the bat's wing since both derive from the vertebrate forelimb, as is recognizable by the shared arm bones and five phalanges. The wings of insects, on the other hand, are analogous to those of bats. As products of convergent evolution, they serve the same function but have a different origin.

This is not the case. Nature abounds with illustrations to the contrary. One that I know firsthand is a pair-bonding Amazonian cichlid, the discus fish, that has achieved the equivalent of mammalian nursing. Once the fry have absorbed the egg yolk, they gather along the flanks of Mom and Dad to nibble mucus off their bodies. The breeding pair secretes extra mucus for this purpose. The young enjoy both nutrition and protection for about a month until they are "weaned" by parents who now turn away each time they approach.[13] No one would use these fish to make a point about the complexity or simplicity of mammalian nursing for the obvious reason that the mechanisms are radically different. All that they share is the function of feeding and raising the young. Mechanism and function are the eternal yin and yang of biology: they interact and intertwine, yet there is no greater sin than confusing the two.

To understand how evolution works its magic across the evolutionary tree, we often invoke the twin concepts of *homology* and *analogy*. Homology refers to shared traits derived from a common ancestor. Thus, the human hand is homologous with the wing of a bat, since both derive from an ancestral forelimb and carry the exact same number of bones to prove it. Analogies, on the other hand, arise when distant animals independently evolve in the same direction, known as *convergent evolution*. The parental care of the discus fish is analogous to mammalian nursing but certainly not homologous, since fish and mammals do not share an ancestor that did the same. Another example is how dolphins, ichthyosaurs (extinct marine reptiles), and fish all have strikingly similar shapes owing to an environment in which a streamlined body with fins serves speed and maneuverability. Since dolphins, ichthyosaurs, and fish did not share an aquatic ancestor, their shapes are analogous. We can apply the same line of thought to behavior. The sensitivity to faces in wasps and primates came about independently, as a striking analogy, based on the need to recognize individual group mates.

Convergent evolution is incredibly powerful. It has equipped both bats and whales with echolocation, both insects and birds with wings, and both primates and opossums with opposable thumbs. It has also produced spectacularly similar species in distant geographic regions, such as the armored bodies of armadillos and pangolins, the prickly defense of hedgehogs and porcupines, and the predatory weaponry of the Tasmanian tiger and the coyote. There is even a primate, the aye-aye of Madagascar, that looks like E.T. with an extremely elongated middle finger (to tap for hollow spots and extract grubs from wood), a trait that it shares with a small marsupial, the long-fingered triok of New Guinea. These species are genetically miles apart, yet they have evolved the same functional solution. We should not be surprised therefore to find similar cognitive and behavioral traits in species that are eons and continents apart. Cognitive rippling is common precisely because it isn't bound by the evolutionary tree: the same capacity may pop up almost anywhere it is needed. Instead of taking this as an argument against cognitive evolution, as some have done, it perfectly fits the way evolution works through either common descent or adaptation to similar circumstances.

A prime example of convergent evolution is the use of tools.

Redefining Man

As soon as an ape sees something attractive yet out of reach, he starts to cast about for a bodily extension. An apple floats in the moat around the zoo island: the ape takes one glance at the fruit before racing around in search of a suitable stick or a few stones that he can throw behind it so that it will float toward him. He distances himself from his goal in order to reach it—an illogical thing to do—while carrying a search image of what tool might work best. He is in a hurry, because if he doesn't return fast enough, someone else will beat him to the prize. If, on the other hand, his goal is to eat fresh green leaves from a tree, the required tool is quite different: something sturdy to climb on. He may work for half an hour to drag and roll a heavy loose tree stump in the direction of the one tree on the island that has a low side branch. The whole reason he needs a tool is to get across the electric wire around the tree. Before making the actual attempt, he has figured out that the low branch will come in handy. I have even seen apes check the hot wires with the hair on the back of their wrist, hand bent inward, barely touching it, but enough to know if the power is on. If it is off, obviously no tool will be needed, and the foliage is fair game.

Apes do not just search for tools for specific occasions; they actually fabricate them. When the British anthropologist Kenneth Oakley, in 1957, wrote *Man the Toolmaker*, which claimed that only humans make tools, he was well aware of Köhler's observations of Sultan fitting sticks together. But Oakley refused to count this as tool manufacture, since it was done in reaction to a given situation rather than in anticipation of an imagined future. Even today some scholars dismiss ape tools by stress-

One of the most complex tool skills is the cracking of tough nuts with rocks. A wild female chimpanzee selects an anvil stone and finds a hammer that fits her hand to open a nut, while her son watches and learns. Only by the age of six will he reach adult proficiency.

ing how human technology is embedded in social roles, symbols, production, and education. A chimpanzee cracking nuts with rocks doesn't qualify; nor, I suspect, does a farmer picking his teeth with a twig. One philosopher even felt that since chimpanzees don't *need* their so-called tools, it remains a feeble comparison.[14]

I feel like recalling my know-thy-animal rule here, according to which we can safely dismiss a philosopher who thinks that wild chimpanzees sit there pounding and pounding hard nuts with rocks, an average of thirty-three blows per consumed kernel, for generation after generation, for no good reason at all. During peak season, chimpanzees at some field sites spend close to 20 percent of their waking hours fishing with twigs for termites or cracking nuts between rocks. It is estimated that they gain nine times as many kilocalories of energy from this activity as they put into it.[15] Moreover, the Japanese primatologist Gen Yamakoshi found that nuts serve as fallback foods when the apes' main nutrition—seasonal fruits—is scarce.[16] Another fallback is palm pith, which is obtained through "pestle pounding." High up in a tree, a chimpanzee stands bipedally at the edge of the tree crown, pounding the top with a leaf stalk, thus creating a deep hole from which fiber and sap can be collected. In other words, the survival of chimpanzees is quite dependent on tools.

Ben Beck gave us the best-known definition of tool use, of which the short version goes as follows: "the external deployment of an unattached environmental object to alter more efficiently the form, position, or condition of another object."[17] Though imperfect, this definition has served the field of animal behavior for decades.[18] Tool manufacture can then be defined as the active modification of an unattached object to make it more effective in relation to one's goal. Note that intentionality matters a great deal. Tools are brought in from a distance and modified with a goal in mind, which is the reason traditional learning scenarios, which revolve around accidentally discovered benefits, have such trouble explaining this behavior. If you see a chimpanzee strip the side branches off a twig to make it right for ant fishing, or collect a fistful of fresh leaves and chew them into a spongelike clump to absorb water from a tree hole, it is hard to miss the purposefulness. By making suitable tools out of raw materials, chimpanzees are exhibiting the very behavior that once defined *Homo faber,* man the creator. This is why the British paleontologist Louis Leakey, when he first heard about such behavior from Goodall, wrote her back, "I feel that scientists holding to this definition are faced with three choices: They must accept chimpanzees as man, they must redefine man, or they must redefine tools."[19]

After the many observations of chimpanzee tool use in captivity, seeing tool use in the wild by the same species did perhaps not come as a surprise, yet its discovery was crucial since it could not be explained away by human influence. Moreover, wild chimps not only use and make tools, but they learn from one another, which allows them to refine their tools over generations. The result is more sophisticated than anything we know in zoo chimps. A good example are the *tool-*

kits, which can be so complex that it is hard to imagine that they were invented in a single step. A typical one was found by the American primatologist Crickette Sanz in the Goualougo Triangle, Republic of Congo, where a chimpanzee may arrive with two different sticks at a particular open spot in the forest. It is always the same combination: one is a stout woody sapling of about a meter long, while the other is a flexible slender herb stem. The chimp then proceeds to deliberately drive the first stick into the ground, working it with both hands and feet the way we do with a shovel. Having made a sizable hole to perforate an army ant nest deep under the surface, she pulls out the stick and smells it, then carefully inserts her second tool. The flexible stem captures bite-happy insects that she pulls up and eats, dipping regularly into the nest below. Apes often climb off the ground, moving onto tree buttresses, to avoid the nasty bites of colony defenders. Sanz collected more than one thousand such tools, which shows how common the perforator-dipping combination is.[20]

More elaborate toolkits are known for chimpanzees in Gabon hunting for honey. In yet another dangerous activity, these chimps raid bee nests using a five-piece toolkit, which includes a pounder (a heavy stick to break open the hive's entrance), a perforator (a stick to perforate the ground to get to the honey chamber), an enlarger (to enlarge an opening through sideways action), a collector (a stick with a frayed end to dip into honey and slurp it off), and swabs (strips of bark to scoop up honey).[21] This tool use is complicated since the tools are prepared and carried to the hive before most of the work begins, and they will need to be kept nearby until the chimp is forced to quit due to aggressive bees. Their use takes foresight and planning of sequential steps, exactly the sort of organization of activities often emphasized for our human ancestors. At one level chimpanzee tool use may seem primitive, as it is based on sticks and stones, but on another level it is extremely advanced.[22] Sticks and stones are all they have in the forest, and we should keep in mind that also for the Bushmen the most ubiquitous instrument is the digging stick (a sharpened stick to break open anthills and dig up roots). The tool use of wild chimpanzees by far exceeds what was ever held possible.

Chimpanzees use between fifteen and twenty-five different tools per community, and the precise tools vary with cultural and ecological circumstances. One savanna community, for example, uses pointed sticks to hunt. This came as a shock, since hunting weapons were thought to be another uniquely human advance. The chimpanzees jab their "spears" into a tree cavity to kill a sleeping bush baby, a small primate that serves as a protein source for female apes unable to run down monkeys the way males do.[23] It is also well known that chimpanzee communities in West Africa crack nuts with stones, a behavior unheard of in East African communities. Human novices have trouble cracking the same tough nuts, partly because they do not have the same muscle strength as an adult chimpanzee, but also because they lack the required coordination. It takes years of practice to place one of the hardest nuts in the world on a level surface, find a good-sized hammer stone, and hit the nut with the right speed while keeping one's fingers out of the way.

The Japanese primatologist Tetsuro Matsuzawa tracked the development of this skill at the "factory," an open space where apes bring their nuts to anvil stones and fill the jungle with a steady rhythm of banging noise. Youngsters hang around the hardworking adults, occasionally pilfering kernels from their mothers. This way they learn the taste of nuts as well as the connection with stones. They make hundreds of futile attempts, hitting the nuts with their hands and feet, or aimlessly pushing nuts and stones around. That they still learn the skill is a great testament to the irrelevance of reinforcement, because none of these activities is ever rewarded until, by about three years of age, the juvenile starts to coordinate to the point that a nut is occasionally cracked. It is only by the age of six or seven that their skill reaches adult level.[24]

When it comes to tool use, chimps always catch the limelight, but there are three other great apes—bonobos, gorillas, and orangutans—that, together with chimps, us, and the gibbons, make up the Hominoid family. Not to be confused with monkeys, Hominoids are large, flat-chested primates without tails. Within this family, we are closest to chimps and bonobos, both of which are genetically nearly identical to us. Naturally, there is heated debate about what the minuscule-sounding 1.2 percent DNA difference between us and them exactly means, but that we are close family is not in doubt. In captivity, the orangutan is an absolute master tool user, dexterous enough to tie knots into loose shoelaces, and to construct instruments. One young male was seen to join three sticks, which he had first sharpened, into two tubes to build a five-section pole to knock down suspended food.[25] Being notorious escape artists, orangs may dismantle their cage so patiently, from day to day and week to week, while keeping dislodged screws and bolts out of sight, that keepers fail to notice what they are doing until it's too late. In contrast, until recently all we knew about wild orangs was that they sometimes scratched their butt with a stick or held a leafy branch over their head during rain. How could a species that is so talented offer so little evidence of tool use in the wild? The inconsistency was resolved when, in 1999, the tool technology of orangutans in a Sumatran peat swamp came to light. These orangs extract honey from bee nests with twigs and use short sticks to remove the seeds embedded in the stinging hairs of *neesia* fruits.[26]

The other ape species, too, are perfectly capable of tool use, and we have already laid to rest the view that gibbons lack this capacity.[27] But reports from the wild remain meager to nonexistent, sometimes suggesting that only chimps are proficient tool users. We see glimpses, such as when gorillas preventively disarm poacher snares, which requires a grasp of basic mechanics, or traverse deep water. When elephants had dug a new water hole in a swampy forest in the Republic of Congo, the German primatologist Thomas Breuer saw a female gorilla, Leah, try to wade across. She stopped when she was waist-deep into it, however—apes hate swimming. Leah returned to shore to pick up a long branch to gauge the water's depth. Feeling around with her stick, she walked bipedally far into the pool before retracing her steps to return to her wailing infant. This example highlights the shortcomings of Beck's classical definition, because even though Leah's stick altered neither anything in the environment nor her own position, it did serve as a tool.[28]

Chimpanzees are recognized as the most versatile primate tool user apart from us, but this heralded position has been challenged. The challenge did not come from any Hominoid but from a small South American monkey. Brown capuchin monkeys have been known for centuries as organ-grinders and more recently as trained helping hands for quadriplegics. They are extremely manipulative and particularly good at tasks that tap into their tendency to smash and bang things. Having had a colony of these monkeys for decades, I know that almost anything you hand them (a piece of carrot, an onion) is going to be pounded to mush on the floor or against the wall. In the wild, they pound oysters for a long time until the mollusk relaxes its muscle so they can pry it open. During the fall, our monkeys in Atlanta collected so many fallen hickory nuts from nearby trees that we'd hear frantic banging sounds the whole day in our office adjacent to the monkey area. It was a happy sound, because capuchins seem to be in their best mood when they are doing things. Not only did they try to break open the nuts, they also employed hard objects (a plastic toy, a block of wood) to smash them with. About half the members of one group learned to do so, whereas the second group never invented the technique despite having the same nuts and tools. This group obviously consumed fewer nuts.

Capuchins' natural predisposition to be persistent pounders sets them up for nut cracking in the field. A Spanish naturalist first reported it five centuries ago, and more recently an international team of scientists found dozens of cracking sites in the Tietê Ecological Park and other sites in Brazil.[29] At one site, capuchins eat the pulp of a large fruit, after which they drop its seeds to the ground. They return a couple of days later to collect those seeds, which by then have dried out and are often infested with larvae, which the monkeys are fond of. Traveling with the seeds stuffed in their hands, mouth, and (prehensile) tail in search of a hard surface, such as a large rock, the monkeys would get a smaller stone to pound the seeds with. While these stones are about the same size as those used by chimps, the capuchins are only about the size of a small cat, so their hammers weigh about one-third of their body! Literally acting as heavy equipment operators, they lift them high above their heads to get a good hit. When the tough seeds are cracked, the larvae are there for the picking.[30]

Capuchin nut cracking thoroughly upset the evolutionary narrative that had been woven around humans and apes. According to this story, we are not the only ones who knew a Stone Age: our closest relatives still live in one. To stress this point, a "percussive stone technology" site (including stone assemblies and the remains of smashed nuts) was excavated in a tropical forest in Ivory Coast, where chimpanzees must have been opening nuts for at least four thousand years.[31] These discoveries led to a human-ape lithic culture story that fit together nicely, tying us to our close relatives.

This is why the discovery of similar behavior in a more distant relative, such as the capuchin monkey—equipped with tails by which they can hang!—was met with surprise and initial grumbling. The monkeys didn't fit. The more we learned, however, the more the nut cracking by capuchins in Brazil began to resemble that of chimpanzees in West Africa. Yet capuchins belong to the neotropical monkeys, a distant group that split off 30 to 40 million years ago from the rest of the primate order. Perhaps the similar tool use was a case of convergent evolution, since both chimps and capuchins are extractive foragers. They break things open, destroy outer shells, and smash things to pulp in order to eat, which might be the context in which their high intelligence evolved. On the other hand, since both are large-brained primates with binocular vision and manipulative hands, there is an undeniably evolutionary connection. The distinction between homology and analogy is not always as clear as we'd like it to be.

To complicate matters, the tool use of capuchins and chimpanzees may not be cognitively at the same level. Over many years of working with both species, I have formed a distinct impression of how they go about their business, which I'll offer here in everyday language. Chimpanzees, like all the apes, think before they act. The most deliberate ape is perhaps the orangutan, but chimps and bonobos, despite their emotional excitability, also judge a situation before tackling it, weighing the effect of their actions. They often find solutions in their heads rather than having to try things out. Sometimes we see a combination of both, as when they start acting on a plan before it is completely formed, which is of course not unusual in our species either. In contrast, the capuchin monkey is a frenzied trial-and-error machine. These monkeys are hyperactive, hypermanipulative, and afraid of nothing. They try out a great variety of manipulations and possibilities, and once they discover something that works, they instantly learn from it. They don't mind making tons of mistakes and rarely give up. There is not much pondering and thinking behind their behavior: they are overwhelmingly action-driven. Even if these monkeys often end up with the same solutions as the apes, they seem to get there in an entirely different way.

While all this may be a gross simplification, it is not without experimental support. An Italian primatologist, Elisabetta Visalberghi, has spent a lifetime studying the tool use of brown capuchins at her facility adjoining the Rome Zoo. In one illuminating experiment, a monkey faced a horizontal transparent tube with a peanut visible in the center. The plastic tube was mounted so that the peanut was at monkey eye level. The monkey couldn't get to it, though, since the tube was too narrow and long. Many objects were available to push the food out, ranging from the most suitable (a long stick) to the least (short sticks, soft flexible rubber). The capuchins made an astonishing number of errors, such as hitting the tube with the stick, vigorously shaking the tube, pushing the wrong material into one end, or pushing short sticks into both ends so

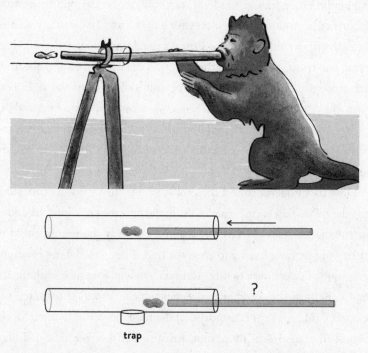

A brown capuchin monkey (top) inserts a long stick into a transparent tube to push out a peanut. Placed in a regular tube, the peanut may be pushed in either direction to solve the problem. The trap-tube (bottom), by contrast, requires the peanut to be pushed in only one direction, otherwise it will drop into the trap and be lost to the monkey. Monkeys can learn to avoid the trap after many errors, but apes show cause-effect understanding and recognize the solution right away.

that the peanut couldn't budge. The monkeys learned over time, however, and began to prefer the long stick.

At this point, Visalberghi added an ingenious twist by making a hole in the tube. Now it suddenly mattered which way the peanut was pushed. Pushed toward the hole, the peanut would fall into a small plastic container and be lost to the monkey. Would capuchin monkeys understand the need to stay clear of the trap, and would they do so right away or only after many failed attempts?

Handing four monkeys a long stick to work on the trap tube, three performed at random, being successful half the time, which they seemed perfectly happy with. But not Roberta, a slender young female, who kept trying and trying. She'd push the stick into the left end of the tube, then race around to see how it and the peanut looked from the right end. Then she'd switch sides, inserting the stick into the right end, only to race around to peek into the tube from the left. She kept going back and forth, sometimes failing, sometimes succeeding, but in the end becoming quite successful.

How had Roberta solved the problem? The investigators concluded that she followed a simple rule of thumb: insert the stick into the end of the tube farthest away from the reward. This way the peanut could be pushed out without having to cross the trap. They tested it out in several ways, one of which was to offer Roberta a new plastic tube without any trap at all. Now she could push the stick whichever way she wanted and be successful. She kept racing around the tube, however, looking for the longest distance from the peanut, insisting on the rule that had been the key to her success. Since Roberta acted as if the trap were still there, she clearly had not paid much attention to how it worked. Visalberghi concluded that monkeys are able to solve the trap-tube task without actually understanding it.[32]

This task may look simple, yet is harder than it seems: human children solve it reliably only when they are over three years old. Testing five chimpanzees on the same problem, two of them grasped the cause-effect relation and learned to specifically avoid the trap.[33] While Roberta had merely learned which actions led to success, the apes recognized how the trap worked. They were representing the connections between actions, tools, and outcomes in their heads. This is known as a *representational* mental strategy, which allows solutions before action. This difference may seem minor, since both monkeys and apes solved the problem, but it is actually huge. The level at which apes understand the purpose of tools affords them incredible flexibility. The richness of their technology, the toolkits, and the frequent toolmaking all prove that higher cognition helps. The American primate expert William Mason concluded in the 1970s that evolution has endowed the Hominoids' with a cognition that sets them apart from the other primates, so that an ape is best described as a thinking being.

> The ape structures the world in which it lives, giving order and meaning to its environment, which is clearly reflected in its actions. It is not very illuminating, perhaps, to describe a chimpanzee as "figuring out" how to proceed, while it sits and stares at the problem before it. Certainly such an assertion lacks originality, as well as precision. But we cannot escape the inference that some such process is at work, and that it has a significant effect on the ape's performance. It seems better to be vaguely correct than positively wrong.[34]

Here Come the Crows!

I first encountered the tube task during a visit to Jigokudani Monkey Park, in Japan, in one of the world's coldest habitats with native primates. Tourist guides use the task to demonstrate monkey intelligence. At the feeding site next to the river, which attracts snow monkeys from the surrounding montane forest, a horizontal transparent tube was baited with a piece of sweet potato. Rather than wielding a stick like the capuchins, one female snow monkey pushed her small infant into the tube while firmly holding on to its tail. The baby crawled toward the food and grabbed it, only to be quickly withdrawn by its loving mom, who pried the prize from its resistant clutch. Another female collected rocks to throw into one end of the tube, so that the food came out the other end.

These are macaques, monkeys much closer to us than capuchins. The most spectacular evidence for macaque tool use has been collected by Michael Gumert, an American primatologist. On Piak Nam Yai

Island off the coast of Thailand, Gumert found an entire population of long-tailed macaques using stone tools. I am very familiar with this species, having done my dissertation on them. Also known as crab-eating macaques, these smart monkeys are rumored to hang their long tails in water to pull up crabs. I have seen them use their tail almost like a stick to obtain food. Unable to control it as South American primates do—a macaque's tail is nonprehensile—they grab their tail with one hand and swap food from outside to inside their cage with it.

Manipulating one's own body appendage is yet another example that stretches the definition of tool use, but there is no doubt that what Gumert discovered is a well-developed technology. His monkeys on the coast collect stones everyday for two purposes. Bigger stones serve as hammers to pound oysters with blunt force until they break open, revealing a delicious rich food source. Smaller stones are used rather like axes, applying a precision grip and more rapid movements, in order to dislodge shellfish from rocks. During the few hours of ebb tide, both food and tools are abundant, an ideal situation for the invention of this seafood technology. It is testimony to the generalized intelligence of primates, because obviously they evolved in the trees, eating fruits and leaves, but here they were surviving on the beach. After humans, chimps, and capuchins, a fourth primate has entered the Stone Age.[35]

Beyond the primates, however, there are quite a few tool-using mammals and birds. Coastal Californians can watch their own floating technology every day among the kelp. The popular furry sea otter swims on his back while using both front paws to smash shellfish against a stone anvil on his chest. He also hammers abalones with a large rock to dislodge them, taking multiple dives to finish this underwater job. A close relative of the otter may possess even more spectacular talents. The honey badger is the star of a viral YouTube video full of expletives to indicate how "badass" this Chuck Norris of the animal kingdom is. The species is even featured on T-shirts emblazoned with "Honey badger don't care." This so-called badger is a small carnivore, which actually—

like the otter—belongs to the weasel family. While I know of no official reports about their skills, a recent PBS documentary features a rescued honey badger named Stoffel who invents multiple ways to escape from his enclosure at a South African rehabilitation center.[36] Assuming that what we see is not a trained trick, he outwits his human caretakers at every turn and displays the sort of insight for his Houdini act that one might expect from an ape, not a honey badger. The documentary shows Stoffel leaning a rake against the wall and claims that he once piled up large stones against it to escape. After all the stones were removed from his enclosure, he apparently constructed a heap of mud balls for the same purpose.

Even though all this is most impressive and begs for further investigation, the greatest challenge to the supremacy of primates has come, not from other mammals, but from a flock of squawking and cawing birds that landed right in the midst of the tool debate. They caused about as much mayhem as they did in that Hitchcock movie.

During the quiet hours in his pet store, my grandfather patiently trained goldfinches to pull a string. This particular finch is known in Dutch as a *puttertje*, a name that refers to the drawing of water from a well. Males that could both sing and draw would fetch a high price. For centuries, these little colorful birds were kept in homes with a chain around their leg, pulling a thimble up from a glass so as to fetch their own drinking water. One such finch is featured in the seventeenth-century Dutch painting central to Donna Tartt's novel *The Goldfinch*. Of course, we don't keep these birds anymore, at least not in this cruel fashion, but their traditional trick is very similar to the one that, in 2002, gave us Betty the crow.

In an aviary at Oxford University, Betty was trying to pull a little bucket out of a transparent vertical pipe. In the bucket was a small piece of meat, and next to the pipe were two tools for her to choose from. One was a straight wire, the other a hooked one. Only with the latter could Betty get a hold of the bucket's handle. After her companion

Inspired by an Aesop fable, crows have been tested to see if they will throw stones into a tube filled with water to bring floating rewards within reach. They do.

stole the hooked wire, however, she faced the task with an inappropriate tool. Undeterred, Betty used her beak to bend the straight wire into a hook so as to pull the bucket from the tube. This remarkable feat was a mere anecdote until perceptive scientists systematically investigated it with new tools. In subsequent tests, Betty received only straight wires, which she kept subjecting to her remarkable bending act.[37] Apart from dispelling the "birdbrain" notion with which birds are unfairly saddled, Betty achieved instant fame by giving us the first laboratory proof of toolmaking outside the primate order. I add "laboratory," because Betty's species in the wild, in the Southwest Pacific, was already known for tool crafting. New Caledonian crows spontaneously modify branches until they have a little wooden hook to fish grubs out of crevices.[38]

The Ancient Greek poet Aesop may have had an inkling of these talents given his fable *The Crow and the Pitcher*. "A Crow, half-dead with thirst," so the fabulist went, "came upon a Pitcher." There wasn't enough water in the pitcher for the crow to be able to drink it. He tried to reach in with his beak, but the water level was too low. "Then a thought came to him," as Aesop put it, "and he took a pebble and dropped it into the Pitcher." Many more pebbles followed until the water had risen enough for a drink. It seems an unlikely feat, but it has now been replicated in the lab. The first was an experiment on rooks, a corvid that in the wild does not use any tools. The rooks were presented with a vertical water-filled tube with a floating mealworm just out of reach. The water level would have to be raised if the rook were to reach the delicacy. The same experiment was carried out with New Caledonian crows, known as real tool experts. True to the dictum that necessity is the mother of invention, and confirming Aesop's story several millennia later, both crow species successfully solved the floating worm puzzle by using pebbles to raise the water level in the tube.[39]

Let me add some caution, though, because it is unclear how insightful this solution was. For one thing, all the birds had been pretrained using a slightly different task. They had received ample rewards for plunging stones into a tube. Moreover, while they were facing the tube with the mealworm, stones had been conveniently placed right next to it. The experimental setup strongly suggested the solution, therefore. Imagine that Köhler had taught his chimps to stack boxes! We would never have heard of him, as it would have undermined any claims of insight. In the course of testing, the crows did learn that large stones work better than small ones, and that there is no point dropping stones into a pipe filled with sawdust. Rather than working these answers out in their minds, however, it may have been a matter of fast learning. Perhaps they noticed that adding stones to water brought the mealworm closer, which led them to persist.[40]

When we recently presented our chimpanzees with a floating peanut task, a female named Liza solved it right away, adding water to a plastic tube. After some vigorous but ineffective kicking and shaking of the tube, Liza abruptly turned around, went to the drinker to fill her mouth, and returned to the tube to add water. She made several more trips to the drinker before she got the peanut at the right level to reach it with her fingers. Other chimps were less successful, but one female tried to pee into the tube! She had the right idea even though the execution was flawed. I have known Liza all her life and am sure that this problem was brand new to her.

Our experiment was inspired by a floating peanut task conducted on a large number of orangutans and chimpanzees, a subset of which cracked the puzzle at first sight.[41] This is especially remarkable, since—unlike the crows—the apes had no pretraining; nor did they find any tools nearby. Rather, they must have conjured the effectiveness of water in their heads before going out of their way to collect it. Water doesn't even look like a tool. How hard this task is became clear from tests on children, many of which never found the solution. Only 58 percent of eight-year-olds came up with it, and only 8 percent of four-year-olds. Most children frantically try to reach the prize with their fingers, then give up.[42]

These studies have set up a friendly rivalry between primate chauvinists and corvid aficionados. I sometimes teasingly accuse the latter of "ape envy," because in every publication they draw a contrast with the primates, saying the corvids are either doing better or at least equally well. Calling their birds "feathered apes," they make outrageous claims such as "The only credible evidence of technological evolution in nonhumans to date comes from New Caledonian crows."[43] Primatologists, on the other hand, wonder how generalizable corvid tool skills are, and if "feathered monkeys" isn't a better moniker for the birds. Are crows one-trick ponies, like the clam-smashing otters or the Egyptian vultures that throw rocks at ostrich eggs? Or do they have the intelligence to take on a broad array of problems?[44] This issue is far from settled, because even though ape intelligence has been studied for over a century, corvid tool studies have come up only in the last decade.

An intriguing new entry is the use of metatools by New Caledonian crows. A crow is presented with a piece of meat that it can retrieve only by using a long stick, but this stick is behind bars wide enough for the crow's beak but not its head. The crow is unable to reach the tool. In a nearby box, however, lies a short stick suitable for retrieving the long one. To solve this problem, the right order is to pick up the short stick, use it to fetch the long one, and then apply the latter stick for the meat. The crow needs to understand that tools can be used on nonfood objects and to take steps in the right order. Alex Taylor and coworkers used wild New Caledonian crows on Maré Island, placed temporarily in an aviary. They tested seven crows, all of which managed metatool use; three followed the right sequence on the first attempt.[45] Presently, Taylor is trying out tasks with even more steps, and the crows are keeping up with the challenge. This is most impressive, and considerably better than monkeys, which have trouble with stepwise tasks.

Given the evolutionary gulf between primates and corvids, and the many ancestral species of mammals and birds in between that don't use tools, we are dealing with a typical example of convergent evolution. Independently, both taxonomic groups must have faced a need for complex manipulations of items in their environment, or other challenges that stimulated brain growth, which led them to evolve strikingly similar cognitive skills.[46] The arrival of corvids on the scene illustrates how discoveries of mental life ripple across the animal kingdom, a process best summarized by what I'll call my cognitive ripple rule: *Every cognitive capacity that we discover is going to be older and more widespread than initially thought.* This is rapidly becoming one of the best-supported tenets of evolutionary cognition.

As a case in point, we now have evidence of tool use outside mammals and birds. Primates and corvids may well show the most sophisticated use of technology, but what to think of partially submerged crocodiles and alligators balancing large sticks on their snouts? Crocodilians do so especially in pools and swamps near rookeries during the nesting season, when herons and other wading birds are in desperate need of sticks and twigs. You can imagine the scene: a heron lands on a log in the water from which it wants to pick up an attractive branch, but suddenly the log comes to life and grabs the bird. Perhaps crocs initially learn that birds land on them when branches float nearby and then extend this association by making sure to be near branches when herons are nesting. From there, it may be a small step to cover oneself with objects that attract birds. The problem with this idea, however, is that there are actually very few free-floating branches and twigs around. There is too much demand for them. Is it possible that the crocs—which the scientists lament are historically taken to be "lethargic, stupid, and boring"—bring their stick-lures with them from far away? This would be another spectacular cognitive ripple, one that extends deliberate tool use to the reptiles.[47]

The final example, which may again stretch the definition of a tool, concerns the veined octopus in the seas around Indonesia. Here we are dealing with an invertebrate: a mollusk! It has been seen collecting coconut shells. Since octopuses are a favorite food of many predators, camouflage is one of their main goals in life. Initially, the coconut shells yield no benefit, however, because they have to be transported, which only draws unwanted attention. Stretching its arms into rigid limbs, the octopus tiptoes over the sea floor while holding its prize in some of its other arms. Awkwardly walking to a safe lair, it can then use the shells to hide underneath.[48] A mollusk collecting tools for future protection, however simple, goes to show how far we have come since the days when technology was thought to be the defining characteristic of our species.

4 | TALK TO ME

Speak and I shall baptize thee!

—French Bishop to a chimpanzee, early 1700s[1]

We associate research in the natural habitat with sacrifice and brav-
ery, since fieldworkers must tackle the unpleasant and dangerous creatures of the tropical rainforest, from bloodsucking leeches to predators
and snakes. By contrast, students of captive animals are thought to
have it easy. But we sometimes forget how much courage it takes to
defend one's ideas in the face of staunch opposition. Most of the time this
occurs just among academics, which is disagreeable rather than hazard-
ous, but Nadia Kohts faced lethal risks. Her full name was Nadezhda
Nikolaevna Ladygina-Kohts, and she lived and worked early last century in the shadow of the Kremlin. Under the sinister influence of the
would-be geneticist Trofim Lysenko, Joseph Stalin had many a brilliant
Russian biologist either shot or sent to the Gulag for thinking the wrong
thoughts. Lysenko believed that plants and animals pass on traits gained
during their lifetime. The names of those who disagreed with him
became unmentionable, and entire research institutes were closed down.

It was in this oppressive climate that Kohts, with her husband Alexander Fiodorovich Kohts—founding director of Moscow's State Dar-

win Museum—set out to study ape facial expressions, inspired by *The Expression of the Emotions in Man and Animals* by that bourgeois Englishman, Charles Darwin. Lysenko was distinctly ambivalent about Darwin's theory, some of which he labeled "reactionary." Staying out of trouble became a major preoccupation of the Kohtses, who hid documents and data among their taxidermy collection in the museum basement. They wisely put a large statue of the French biologist Jean-Baptiste Lamarck—famous proponent of the inheritance of acquired characteristics—at the museum entrance.

Kohts published in French, German, and most of all her native Russian. She wrote seven books, of which only one was translated into English, long after its appearance in 1935. The English version of *Infant Chimpanzee and Human Child*, edited by me, appeared in 2002. The book compares the emotional life and intelligence of a young chimpanzee, Joni, with that of Kohts's little son, Roody. Kohts studied Joni's reactions to pictures of chimpanzees and other animals, and to his own mirror image. Even though Joni was probably too young to recognize himself, Kohts describes how he would entertain himself in front of his reflection by pulling weird faces and sticking out his tongue.[2]

Kohts is little known compared to Wolfgang Köhler, who conducted his groundbreaking ape research from 1912 through 1920. I wonder what she knew about it while working in Moscow from 1913 until Joni's premature death in 1916. While Köhler is widely recognized as a pioneer of evolutionary cognition, pictures of Kohts's work leave little doubt that she was on exactly the same track. One of the museum's glass cases features Joni's mounted body surrounded by ladders and tools, including sticks that fit into each other. Was Kohts overlooked by science due to her gender? Or was it her language?

I learned about her from the writings of Robert Yerkes, who came to Moscow to discuss her projects through an interpreter. In his books, Yerkes described Kohts's work with the greatest admiration. There is a good chance, for example, that Kohts invented the matching-to-sample

Nadia Ladygina-Kohts was a pioneer in animal cognition, who studied not only primates but also parrots, such as this macaw. Working in Moscow at around the same time that Köhler conducted his research, she remains far less known.

(MTS) paradigm, a staple of modern cognitive neuroscience. MTS is nowadays being applied to both humans and animals in countless laboratories. Kohts would hold up an object for Joni, then hide it among other objects in a sack and let him feel around to find the first one. The test involved two modalities—vision and touch—demanding that Joni make a choice based on his memory of the previously seen model.

My own fascination with this unsung hero's work took me to Moscow, too. I received a behind-the-scenes tour of the museum, where I leafed through private picture albums. Kohts was (and is) much beloved in her country, where she is widely recognized as the great scientist that she was. My biggest surprise was to learn that she owned at least three large parrots. Pictures show her accepting an object handed to her by a cockatoo, and her holding out a tray with three cups toward a macaw. The parrots would sit opposite her, on a table, while Kohts held a small food reward in one hand and a pencil in the other, scoring their choices as she tested their ability to discriminate among objects. I checked with our contemporary expert on Psittaciformes, the American psychologist

Irene Pepperberg, but she had never heard of Kohts's parrot studies. I doubt that anyone in the West ever suspected that bird cognition, too, was studied in Russia well before it became more widely known.

Alex the Parrot

I first met Alex, the African gray that Irene raised and studied for three decades, on visits to her department from a nearby university. Irene had bought the bird in a pet store, in 1977, and was setting up an ambitious project that would open the public eye to the avian mind. It ended up paving the way for all subsequent studies of bird intelligence, because until then the general opinion had been that bird brains simply don't support advanced cognition. Due to their lack of much of anything that looks like a mammalian cortex, birds were viewed as well endowed with instincts yet poor at learning, let alone thinking. Despite the fact that their brains can be quite sizable—the African gray's is the size of a shelled walnut, with a large area that functions like a cerebral cortex—and that their natural behavior offers ample reason to question the low opinion of them, the different brain organization of birds has been held against them.

Having myself kept and studied jackdaws—members of that other large-brained bird family, the corvids—I have never had any doubt about their behavioral flexibility. On walks through the park, my birds would tease dogs by flying right in front of their heads, just out of reach of their snapping mouths, to the surprise and chagrin of the dog owners. Indoors, they would play object hiding with me: I would hide a small item, such as a cork, under a pillow or behind a flower pot, while they would try to find it, or vice versa. This game relied on the well-known food-caching talents of crows and jays but also suggested *object permanence*: the understanding that an object continues to exist even after it has disappeared from view. The extreme playfulness of my jackdaws hinted, as it does with animals in general, at high intelligence and the thrill of a challenge. Visiting Irene, I was quite prepared to be impressed by a bird, therefore, and Alex did not disappoint. Cockily sitting on his perch, he had begun to learn labels for items such as keys, triangles, and squares, saying "key," "three-corner," or "four-corner" whenever these objects were pointed out.

At first sight, this came across as language learning, but I am not sure this is the right interpretation. Irene didn't claim that Alex's talking amounted to speaking in the linguistic sense. But of course, the labeling of objects is very much part of language, and we should not forget that once upon a time linguists defined language simply as symbolic communication. Only when apes proved capable of such communication did they feel the need to raise the bar and add refinements such as that language requires syntax and recursivity. Language acquisition by animals became a huge topic that drew enormous public interest. It was as if all questions about animal intelligence boiled down to a sort of Turing test: can we, humans, hold a sensible conversation with them? Language is such a marker of humanity that an eighteenth-century French bishop was ready to baptize an ape provided he could speak. It surely was all that science seemed to care about in the 1960s and 1970s, resulting in attempts to talk with dolphins and teach language to a multitude of primates. Some of this attention turned sour, however, when the American psychologist Herbert Terrace, in 1979, published a highly skeptical article about the sign-language capacities of Nim Chimpsky, a chimpanzee named after American linguist Noam Chomsky.[3]

Terrace found Nim a boring conversationalist. The vast majority of his utterances were requests for desirable outcomes, such as food, rather than expressions of thoughts, opinions, or ideas. Terrace's surprise at this was by itself rather surprising, however, given his reliance on operant conditioning. Since this is not how we teach children language, one wonders why it was used for an ape. Having been rewarded thousands of times for hand signals, why wouldn't Nim use these signals to obtain rewards? He simply did what he was taught. As a result of this project, however, the voices pro and contra animal language were getting louder by the day. To find a bird voice among this cacophony threw many people off, because while apes obviously don't talk, Alex carefully pronounced every word. Superficially, his behavior resembled language more than that of any other animal, even if there was little agreement about what it actually meant.

Irene's choice of species was intriguing since Doctor Dolittle, the central character of a series of children's books, owned an African gray, named Polynesia, who taught the good doctor the language of animals. Irene had always been attracted to these stories and as a child already presented her pet budgie with a drawer full of buttons to see how the bird would arrange them.[4] Her work with Alex grew straight out of her early captivation with birds and their taste in colors and shapes. But before discussing her research further, let me briefly dwell on the desire to talk with animals—a desire often expressed by scientists working on animal cognition—as it relates to the deeper connection often assumed between cognition and language.

Oddly enough, this particular desire must have passed me by, because I have never felt it. I am not waiting to hear what my animals have to say about themselves, taking the rather Wittgensteinian position that their message might not be all that enlightening. Even with respect to my fellow humans, I am dubious that language tells us what is going on in their heads. I am surrounded by colleagues who study members of our species by presenting them with questionnaires. They trust the answers they receive and have ways, they assure me, of checking their veracity. But who says that what people tell us about themselves reveals actual emotions and motivations?

This may be true for simple attitudes free from moralizations ("What is your favorite music?"), but it seems almost pointless to ask people about their love life, eating habits, or treatment of others ("Are you pleasant to work with?"). It is far too easy to invent post hoc reasons for one's behavior, to be silent about one's sexual habits, to downplay excessive eating or drinking, or to present oneself as more admirable than one really is. No one is going to admit to murderous thoughts, stinginess, or being a jerk. People lie all the time, so why would they stop in front of a psychologist who writes down everything they say? In one study, female college students reported more sex partners when they were hooked up to a fake lie-detector machine than without it, thus demonstrating that they had been lying before.[5] I am in fact relieved to work with subjects that don't talk. I don't need to worry about the truth of their utterances. Instead of asking them how often they engage in sex, I just count the occasions. I am perfectly happy being an animal watcher.

Now that I think of it, my distrust of language goes even deeper, because I am also unconvinced of its role in the thinking process. I am not sure that I think in words, and I never seem to hear any inner voices. This caused a bit of an embarrassment once at a meeting about the evolution of conscience, when fellow scholars kept referring to an inner voice that tells us what is right and wrong. I am sorry, I said, but I never hear such voices. Am I a man without a conscience, or do I—as the American animal expert Temple Grandin once famously said about herself—think in pictures? Moreover, which language are we talking about? Speaking two languages at home and a third one at work, my thinking must be awfully muddled. Yet I have never noticed any effect, despite the widespread assumption that language is at the root of human thought. In his 1973 presidential address to the American Philosophical Association, tellingly entitled "Thoughtless Brutes," the American philosopher Norman Malcolm stated that "the relationship between language and thought must be so close that it is really senseless to conjecture that people may *not* have thoughts, and also senseless to conjecture that animals *may* have thoughts."[6]

Since we routinely express ideas and feelings in language, we may be forgiven for assigning a role to it, but isn't it remarkable how often we struggle to find our words? It's not that we don't know what we thought or felt, but we just can't put our verbal finger on it. This would of course be wholly unnecessary if thoughts and feelings were linguistic products to begin with. In that case, we'd expect a waterfall of words! It is now widely accepted that, even though language assists human thinking by providing categories and concepts, it is not the stuff of thought. We don't actually need language in order to think. The Swiss pioneer of cognitive development, Jean Piaget, most certainly was not ready to deny thought to preverbal children, which is why he declared cognition to be independent of language. With animals, the situation is similar. As the chief architect of the modern concept of mind, the American philosopher Jerry Fodor, put it: "The obvious (and I should have thought sufficient) refutation of the claim that natural languages are the medium of thought is that there are non-verbal organisms that think."[7]

What irony: we have traveled all the way from the absence of language as an argument against thought in other species to the position that the manifest thinking by nonlinguistic creatures argues against the importance of language. While I won't complain about this turn of events, it owes a great debt to language studies on animals such as Alex: not so much because these studies demonstrated language per se but because they helped expose animal thought in a format that we easily relate to. We see a sharp-looking bird, who replies when spoken to, pronouncing object names with great accuracy. He faces a tray full of objects, some made of wool, some of wood, some of plastic, representing all colors of the rainbow. He is invited to feel every object with his beak and tongue, and then, after they have all been returned to the tray, he is asked what the two-cornered blue object is made of. By correctly answering "wool," he combines his knowledge of color, shape, and material with his memory of what this particular item felt like. Or he sees two keys, one made of green plastic, the other of metal, and is asked "what is different?" He says "color." Asked "which color bigger," he answers "green."[8]

Anyone watching Alex perform, as I did in the early stages of his career, is blown away. Obviously, skeptics tried to ascribe his skills to rote learning, but since the stimuli changed all the time as did the questions asked, it is hard to see how he could have performed at this level based on stock answers. He would have needed a gigantic memory to handle all possibilities, so much so that it is in fact simpler to assume, as Irene did, that he had acquired a few basic concepts and was capable of mentally combining them. Furthermore, he didn't need Irene's presence to answer, nor did he even need to see the actual items. In the absence of any corn, he might be asked what color corn is and would say "yellow." Particularly impressive was Alex's ability to distinguish "same" from "different," which required him to compare objects on a variety of dimensions. All these capacities—labeling, comparing, and judging color, shape, and material—were assumed to require language at the time that Alex began his training. It was an aggravating struggle for Irene to convince the world of his skills, especially since skepticism with respect to birds ran so much deeper than it ever was for our close relatives, the primates. After years of persistence and solid data, however, she had the satisfaction of seeing Alex turn into a celebrity. Upon his death in 2007, he was honored with obituaries in both *The New York Times* and *The Economist*.

In the meantime some of his relatives had begun to impress as well. Another African gray not only mimicked sounds but added accompanying body movements. He'd say "Ciao" while waving goodbye with a foot or wing, or say "Look at my tongue" while sticking out his tongue, just as his owner had shown him. It remained a puzzle how a bird was able to draw such parallels between the human body and its own.[9] Then there was Figaro, a Goffin's cockatoo who was seen breaking off large splinters from a wooden beam in order to rake in nuts placed outside his aviary. Before Figaro, there had been no reports of toolmaking parrots.[10] It makes me wonder if Kohts ever conducted similar experiments on her cockatoo, macaw, and ara. Given her keen interest in tools and her six untranslated books, I wouldn't be surprised to hear about it one day. There is obviously still much to discover, as also became clear from tests of Alex's counting abilities.

Alex's talents were accidentally revealed while researchers were testing Griffin—a parrot named after Donald Griffin—who was staying in the same room with him. In order to see if Griffin could pair quantities with sounds, they would click twice, to which the right answer would be "two." But when Griffin failed to answer and got two more clicks, Alex, from across the room, chimed in with "four." And after two more clicks, Alex said "six," while Griffin remained mute.[11] Alex was familiar with numbers and could correctly answer the question "what number is green?" after having seen a tray with many objects, including several green ones. But now he was doing addition, and more than that: he was doing it without visual input. Again, adding up numbers was once thought to be language-dependent, but this claim had already begun to wobble a few years back when a chimpanzee succeeded at it.[12]

Irene set out to test Alex's capacities more systematically by placing a few differently sized items (such as pasta pieces) under a cup. She'd lift the cup up for a few seconds in front of Alex, then put it down again. After this she would do the same for a second cup, then a third. The number of items under each cup was small, and sometimes there were none. After this, with only the three cups visible, Alex would be asked "how many total?" Out of ten tests, Alex mentioned the correct total eight times. The two that he missed, he got right the second time he heard the question.[13] And all this in his head, because he couldn't see the actual items.

Unfortunately, this study was broken off by Alex's unexpected death. But by then this diminutive mathematical genius in a grey suit had given us ample evidence that there is more knocking around in a bird's skull than anyone had suspected. Irene concluded that "for far too long, animals in general, and birds in particular, have been denigrated and treated merely as creatures of instinct rather than as sentient beings."[14]

Red Herring

At times, Alex's talking made perfect linguistic sense. For example, once when Irene was fuming about a meeting in her department and walked to the lab with angry steps, Alex told her "Calm down!" No doubt the same expression had in the past been aimed at Alex's own excitable self. Other famous cases include Koko, the sign-language gorilla spontaneously combining the signs for "white" and "tiger" upon seeing a zebra, and Washoe, the chimpanzee pioneer of this entire field, labeling a swan a "water bird."

I am prepared to interpret this as a hint of deeper knowledge, but only after I see more evidence than we have today. It is good to keep in mind that these animals produce hundreds of signs every day and have been studied for decades. We'd need to know more about the ratio between hits and misses among the thousands of utterances recorded. How are these fortuitous combinations different from, say, Paul the octopus (nicknamed Pulpo Paul) who rose to fame after a string of correct predictions during the 2010 World Cup? In the same way that no one assumes that Paul knew much about soccer—he was just a lucky mollusk—we need to compare striking animal utterances with the probability of them coming about by chance. It is hard to evaluate linguistic skills if we never get to see the raw data, such as unedited videotapes, and hear only cherry-picked interpretations by loving caretakers. It also doesn't help that whenever apes produce wrong answers, their interpreters assume that they have a sense of humor, exclaiming "Oh, stop kidding around!" or "You funny gorilla!"[15]

Upon the death of Robin Williams, in 2014, when the whole country was grieving one of the world's funniest men, Koko was said to be mourning, too. It sounded plausible, especially since the Gorilla Foundation, in California, called Williams one of her "closest friends." The problem is that the two of them had met just once, thirteen years before, and that the only evidence of Koko's "somber" reaction was a photo of her sitting with her head down and eyes closed, which was hard to distinguish from a dozing ape. I found the grieving claim to be a huge stretch, not because I doubt that apes have feelings or can grieve, but because it is nearly impossible to gauge an animal's reaction to an event it has not witnessed. While it is entirely possible that Koko's mood was affected by the people around her, this is not the same as grasping what had happened to a member of our species whom she barely knew.

All responses to death and loss thus far observed in apes concern individuals who were truly close (such as mother and offspring, or lifelong friends) and whose corpses the apes were able to see and touch. Mourning triggered by the mere mention of someone's death requires a level of imagination and understanding of mortality that most of us don't assume. It is precisely because of such inflated claims that the whole field of talking apes has fallen into ill repute over the years, and why no new projects of its kind are being initiated. Those that still do exist tend to resort to feel-good stories and publicity stunts to raise funds. There is too much of this going around, and too little hard-nosed science.

You won't often hear me say something like this, but I consider us the only linguistic species. We honestly have no evidence for symbolic communication, equally rich and multifunctional as ours, outside our species. It seems to be our own magic well, something we are exceptionally good at. Other species are very capable of communicating inner processes, such as emotions and intentions, or coordinating actions and plans by means of nonverbal signals, but their communication is neither symbolized nor endlessly flexible like language. For one thing, it is almost entirely restricted to the here and now. A chimpanzee may detect another's emotions in reaction to a particular ongoing situation, but cannot communicate even the simplest information about events displaced in space and time. If I have a black eye, I can explain to you how yesterday I walked into a bar with drunken people . . . and so on. A chimpanzee has no way, after the fact, to explain how an injury came about. Possibly, if his assailant happens to walk by and he barks and screams at him, others will be able to *deduce* the connection between his behavior and the injury—apes are smart enough to put cause and effect together—but this would work only in the other's presence. If his assailant never walks by, there will be no such information transfer.

Countless theories have attempted to identify the benefits that language bestows upon our species and to explain why language may have arisen. In fact, an entire biennial international conference is devoted to exactly this topic, where speakers present more speculations and evolutionary scenarios than you can imagine.[16] I myself take the rather simple view that the first and foremost advantage of language is to transmit information that transcends the here and now. There is great survival value in communication about things that are absent or events that have happened or are about to happen. You can let others know that there is a lion over the hill, or that your neighbors have picked up weapons. This is just one idea out of many, though, and it is true that modern languages are far too complex and elaborate for this limited purpose. They are sophisticated enough to express thoughts and feelings, convey knowledge, develop philosophies, and write poetry and fiction. What an incredibly rich capacity it is: one that seems entirely our own.

But as with so many larger human phenomena, once we break it down into smaller pieces, some of these pieces can be found elsewhere. It is a procedure I have applied myself in my popular books about primate politics, culture, even morality.[17] Critical pieces such as power alliances (politics) and the spreading of habits (culture), as well as empathy and fairness (morality), are detectable outside our species. The same holds for capacities underlying language. Honeybees, for example, accurately signal distant nectar locations to the hive, and monkeys may utter calls in predictable sequences that resemble rudimentary syntax. The most intriguing parallel is perhaps *referential signaling*. Vervet monkeys on the plains of Kenya have distinct alarm calls for a leopard, eagle, or snake. These predator-specific calls constitute a life-saving communication system, because different dangers demand different responses. For example, the right response to a snake alarm is to stand upright in the tall grass and look around, which would be suicidal in case a leopard lurks in the grass.[18] Instead of having special calls, some other monkey species combine the same calls in different ways under different circumstances.[19]

After the primate studies, the usual rippling has added birds to the list of referential signalers. Great tits, for example, have a unique call for snakes, which pose a grave threat as they slither into nests to swallow the young.[20] But whereas these kinds of studies have helped raise the profile of animal communication, some serious doubts have been raised, too, and language parallels have been called a "red herring."[21] Animal calls do not necessarily mean what we think they mean: a critical part of how they function is how listeners interpret them.[22] On top of this, it is good to keep in mind that most animals do not learn their calls the way humans learn words. They are simply born with them. However sophisticated natural animal communication may be, it lacks the symbolic quality and open-ended syntax that lends human language its infinite versatility.

Perhaps hand gestures offer a better parallel, since in the apes they are under voluntary control and often learned. Apes move and wave their hands all the time while communicating, and they have an impressive repertoire of specific gestures such as stretching out an open hand to beg for something, or moving a whole arm over another as a sign of dominance.[23] We share this behavior with them and only them: monkeys have virtually no such gestures.[24] The manual signals of apes are intentional, highly flexible, and used to refine the message of communication. When a chimp holds out his hand to a friend who is eating, he is asking for a share, but when the same chimp is under attack and holds out his hand to a bystander, he is asking for protection. He may even point out his opponent by making angry slapping gestures in his direction. But although gestures are more context-dependent than other signals and greatly enrich communication, comparisons with human language remain a stretch.

Does this mean that all the attempts to find languagelike qualities in animal communication have been a waste of time, including training projects, such as those with Alex, Koko, Washoe, Kanzi, and others? After Terrace's paper, linguists eager to rid their territory of hairy or feathered "intruders" made the fruitlessness of animal research their mantra. They were so contemptuous of it that, at a 1980 conference—the title of which contained the words *Clever Hans*—they called for an official *ban* on any and all attempts to teach animals language.[25] This unsuccessful move was reminiscent of nineteenth-century anti-Darwinists for whom language was the one barrier between brute and man, including the Linguistic Society of Paris, which in 1866 forbade the study of language origins.[26] Such measures reflect intellectual fear rather than curiosity. What are linguists afraid of? They had better pull their heads out of the sand, because no trait, not even our beloved linguistic ability, ever comes about de novo. Nothing evolves all of a sudden, without antecedents. Every new trait taps into existing structures and processes. Thus, Wernicke's area, a part of the brain central to human speech, is recognizable in the great apes, in which it is enlarged on the left side, as it is in us.[27] This obviously raises the question of what this particular brain region was doing in our ancestors before it was recruited for language. There are many such connections, including the FoxP2 gene that affects both human articulated speech and the fine motor control of birdsong.[28] Science increasingly views human speech and birdsong as products of convergent evolution, given that songbirds and humans share at least fifty genes specifically related to vocal learning.[29] No one serious about language evolution will ever be able to get around animal comparisons.

In the meantime, language-inspired studies have dispelled the notion that natural animal communication is purely emotional. We now have a far better grasp of how communication is geared to an audience, provides information about the environment, and relies on interpretation by those receiving the signals. Even if the connection with human language remains contentious, our appreciation of animal communication has greatly benefited from this flurry of research. As for the handful of language-trained animals, they have proven invaluable at showing what their minds are capable of. Since these animals respond to requests and prompts in a way that we find easy to interpret, the results speak to the human imagination and have been instrumental in breaking open the field of animal cognition. When Alex hears a question about the items on his tray, he inspects them carefully and comments on the one that he was asked about. We have no trouble putting ourselves into his shoes, given that we understand both the question and his answer.

I once asked Sue Savage-Rumbaugh, who worked with Kanzi, the bonobo who communicates by pressing symbols on a keyboard, "Would you say that you study language or intelligence, or is there no difference?" She replied:

> There is a difference because we have apes who have no linguistic abilities in the human sense, but who do quite well on cognitive tasks such as solving a maze problem. Language skills can help elaborate and refine cognitive skills, though, because you can tell an ape who is language-trained something that he does not know. This can put a cognitive task on a whole different plane. For example, we have a computer game in which apes put three puzzle pieces together to make different portraits. After having learned this, they get four pieces presented on the screen, and the fourth piece is from a different portrait. When we first did this with Kanzi, he would take the piece of a bunny face and put it together with a piece of my face. He kept trying, but of course it wouldn't fit. Since he understands spoken language so well, I could say to him: "Kanzi, we're not making the bunny, put Sue's face together." As soon as he heard this, he stopped making the bunny, and stuck to the pieces of my face. So, the instructions had an immediate effect.[30]

Since Kanzi lived for years in Atlanta, I met him multiple times and was always impressed by how well he grasped spoken English. What struck me was not his self-produced utterances—which were rather basic, certainly below the level of a three-year-old child—but the way he reacted to those by the people around him. In one videotaped exchange, Sue asks him "Put the key in the refrigerator," while she wears a welding mask to prevent Clever Hans Effects. Kanzi picks up a chain of keys, opens the fridge, and puts the keys into it. Asked to give his doggy a shot, he picks up a plastic syringe and injects it into his stuffed toy dog. Kanzi's passive comprehension is greatly helped by his familiarity with a large number of items and words. This has been tested by playing spoken words to him through headphones while he sits at a table and selects a picture of the object that he hears being mentioned. But that he is excellent at word recognition still doesn't explain why Kanzi appears to understand entire sentences.

Such understanding is something I also know of my own apes despite the fact that none of them have had language training. Georgia is a naughty chimpanzee prone to furtively collecting water from the faucet so as to spray unsuspecting visitors. Once I told her, in Dutch, while pointing a finger at her, that I had seen her. Immediately, she let the water run from her mouth, apparently realizing that there was no point trying to surprise us. But how did she know what I had said? My suspicion is that many apes know a few key words and are highly sensitive to contextual information, such as our tone of voice, glances, and gestures. After all, Georgia had just collected a mouthful of water, and I was giving a range of clues, such as pointing a finger at her and calling her by name. Without necessarily following my exact words, she had the cognitive talent to piece together what I probably meant.

When apes guess correctly, we get the distinct impression that they must have understood everything we said, but their understanding may be more fragmentary. A striking illustration was given by Robert Yerkes after an interaction with Chimpita, a young male chimpanzee:

I was feeding grapes to Chimpita one day and he swallowed the seeds. I told him he must give the seeds to me, for I was afraid they might cause appendicitis, so he gave me all the seeds he had in his mouth and then picked up some from the floor with his lips and his hands. Finally, there were two left between the cage wall and the cement floor which he could not get well with either lips or fingers. I said to him "Chimpita, when I have gone you will eat those seeds." He looked at me as if he asked why I bothered him so much. Then he went into the next cage, looking at me all the while, got a little stick, and with it poked the seeds out of the crack and gave them to me.[31]

It is easy to think that Chimpita must have understood the whole sentence, which is why an astonished Yerkes added, "Such behavior demands careful scientific analysis." But more likely, the ape was following the scientist's body language more closely than we are used to. I regularly have this eerie impression that apes look right through me, perhaps because they are not distracted by language. By directing our attention to what others have to say, we neglect body language compared to animals, for whom it is all they have to go by. It is a skill they employ every day and have refined to the point that they read us like a book. It reminds me of a story by Oliver Sacks about a group of patients in an aphasia ward who were convulsed with laughter during a televised speech by President Ronald Reagan.[32] Incapable of understanding words as such, aphasia patients follow what is being said through facial expressions and body language. They are so attentive to nonverbal cues that they cannot be lied to. Sacks concluded that the president, whose speech seemed perfectly normal to others around, so cunningly combined deceptive words and tone of voice that only the brain-damaged were able to see through it.

The immense effort to find language outside our own species has, ironically, led to a greater appreciation of how special the language capacity is. It is fed by specific learning mechanisms that allow a toddler to linguistically outpace any trained animal. It is in fact an excellent example of biologically prepared learning in our species. Yet this realization by no means invalidates the revelations we owe animal language research. That would be like throwing out the baby with the bathwater. It has given us Alex, Washoe, Kanzi, and other prodigies who have helped put animal cognition on the map. These animals convinced skeptics and the general public alike that there is much more to their behavior than rote learning. One cannot watch a parrot successfully count up items in his head and still believe that the only thing these birds are good at is parroting.

To the Dogs

Each in their own way, Irene Pepperberg and Nadia Kohts navigated treacherous waters. It would be great if everyone were open-minded and purely interested in the evidence, but science is not immune to preconceived notions and fanatically held beliefs. Anyone who forbids the study of language origins must be scared of new ideas, as must anyone whose only answer to Mendelian genetics is state persecution. Like Galileo's colleagues, who refused to peek through his telescope, humans are a strange lot. We have the power to analyze and explore the world around us, yet panic as soon as the evidence threatens to violate our expectations.

This was the situation when science got serious about animal cognition. It was an upsetting time for many. The language studies helped kill the reigning incredulity, even if for reasons other than their original intent. With the cognitive genie out of the bottle, it couldn't be pushed back in, and science began to explore animals through less language-colored glasses. We returned to the ways Kohts, Yerkes, Köhler, and others had conceived their studies, focusing on tools, knowledge of the environment, social relations, insight, foresight, and so on. Many experimental paradigms popular today in studies of cooperation, food sharing, and token exchange go back to research of one century ago.[33] Of course, there remains the problem of how to work with hard-to-control creatures, such as the apes, and how to motivate them. If they haven't grown up around humans, these animals have no clue what our commands mean and don't pay as much attention to us as we'd like them to. They remain essentially wild and hard to engage. Language-trained animals have been so much easier to deal with that one wonders how we might replace them.

In most cases this is impossible, and we'll just have to learn how to test wild or semiwild creatures. But there is one exception, which is an animal intentionally bred by our species to get along with us: the dog. Not so long ago, students of animal behavior shied away from dogs precisely because they were domesticated animals, hence genetically modified and artificial. But science is coming around to the dog, recognizing its advantage for studies on intelligence. For one thing, dog researchers don't need to worry as much about safety or to lock their subjects up in cages. They don't need to feed or maintain their subjects, since they just ask people to drop by at a convenient time with their pets. They compensate the proud owners with a certificate emblazoned with the seal of their university, which confirms their pooch's genius. Most of all, investigators don't face the motivational problems found in most other animals. Dogs eagerly pay attention to us and need little encouragement to work on the tasks that we present to them. No wonder "dognition" is an up-and-coming field.[34] In the meantime, we are also learning more about human perceptions of animals. Did you know, for example, that one quarter of dog owners believe their pets to be smarter than most people?[35] As an added bonus, the dog is a highly empathic and social creature, so that these studies also illuminate animal emotions, an area Darwin was excited about. He often used dogs to illustrate the emotional continuity among species.

With dogs, we even have the prospect of neuroscience at a level that remains out of reach for most other animals. In our own species, we are used to fMRI scans of the brain in order to see what we are afraid of or how much we love each other. Results of these studies are common fare in the news media. Why aren't we doing the same with animals? The reason is that humans are prepared to lie still for many minutes inside a giant magnet, which is the only way to get a good image of their brains. We can ask them questions and show them videos and compare their brain's activity with its resting state. The answers are not always as informative as they are hyped to be, though, because brain imaging often amounts to what I mockingly call *neurogeography*. The typical outcome is a brain map with an area lighted up in yellow or red: it tells us *where* things happen in the brain, but rarely do we hear an explanation of *what* is going on and *why*.[36]

Apart from this limitation, however, the problem that has vexed science is how to gather the same information on animals. Attempts have been made with birds, but they were not awake during the scanning itself. We also have brain scans of immobilized yet awake marmosets. Put in a scanner swaddled like Mongolian babies, these tiny monkeys were exposed to various scents.[37] But for larger primates, such as chimpanzees, to undergo such a procedure—even if it were at all practical, which it is not—would cause so much stress that it would keep them from paying attention to cognitive tasks. We also cannot put them under anesthesia, since this would defeat the whole purpose. The real challenge is to get fully conscious voluntary participation.

To see how this may be done, I descended one day to the basement floor of my own psychology department at Emory University to inspect the new magnet intended for human imaging. One of my colleagues had begun to exploit this fine piece of equipment to achieve a breakthrough with the one animal that can be trained to sit still. Gregory Berns, a neuroscientist, joined me in the waiting room with Eli, a large intact male dog, and Callie, a much smaller spayed female. Callie is the hero of Greg's tale, as she is his own pet, the first dog trained to lie still with her snout in a specially designed holder.

While we waited, the dogs played nicely together in the room, but

Callie in a magnetic resonance scanner. Dogs can be trained to sit still, which permits the study of their cognition through brain imaging, such as fMRI.

when it turned into a fight in which Eli drew a drop of blood, we had to separate them. This was surely different from most human waiting rooms. For Callie, it was the eighth time she had received the mutt-muffs, or foam-filled ear seals that fit like headphones over a dog's head to reduce sound, such as the buzzing of the magnet. It is an important part of the project to get the dogs used to odd noises. Strangely enough, Greg was convinced that this might work after seeing a video of the raid on Osama bin Laden's compound. SEAL Team 6 had a trained dog jump out of a helicopter with an oxygen mask on while strapped to a soldier's chest. If you can train dogs to do this, Greg thought, we certainly should be able to get them used to the magnet's noises. This, together with training them to put their heads in a chinrest, is the secret to the project's success. With lots of little chunks of hotdog, the canines are trained at home so that the chinrest in the magnet is familiar to them and they know what is expected of them.[38]

The frequent rewards pose a bit of a problem, because eating requires jaw movements, which interfere with brain imaging. Via a special dog ladder, Callie ran into the scanner and took her position waiting for the procedure. She was a bit too excited, though, because her tail wagged wildly, adding another source of body movement. Greg's joking that we were looking for the tail-wagging area in the brain was not too far off. Eli needed a bit more encouragement to enter the scanner but was convinced once he saw his familiar chinrest. His owner told me that he is so used to it, and associates it with such good times, that she sometimes finds him sleeping at home with his head inside. He remained still for three minutes, long enough for some good scanning.

Pretrained hand signals tell the dog in the scanner whether a treat is forthcoming. This is how Greg studies activation of their pleasure centers. His goals are rather modest at this point, such as to show that similar cognitive processes in humans and dogs engage similar brain areas. Greg is finding that the prospect of food activates the caudate nucleus in the canine brain in the same way that it does in the brain of businessmen anticipating a monetary bonus.[39] That all mammalian brains operate in essentially the same way has also been found in other domains. Behind these similarities is a much deeper message, of course. Instead of treating mental processes as a black box, as Skinner and his followers had done, we are now prying open the box to reveal a wealth of neural homologies. These show a shared evolutionary background to mental processes and offer a powerful argument against human-animal dualism.

Although this research is still in its infancy, it promises a noninvasive neuroscience of animal cognition and emotion. I felt as if I were at the threshold of a new era, while Eli trotted out of the scanner to lean his head on my knee and let out a deep dog sigh to signal his relief that all had ended well.

5 | THE MEASURE OF ALL THINGS

Ayumu had no time for me while he was working on his computer. He lives with other chimps in an outdoor area at the Primate Research Institute (PRI) of Kyoto University. At any moment, an ape can run into one of several cubicles—like little phone booths—equipped with a computer. The chimp can also leave the cubicle whenever he wants. This way playing computer games is entirely up to them, which guarantees sound motivation. Since the cubicles are transparent and low, I could lean on one to look over Ayumu's shoulder. I watched his incredibly rapid decision making the way I admire my students typing ten times faster than me.

Ayumu, is a young male who, in 2007, put human memory to shame. Trained on a touchscreen, he can recall a series of numbers from 1 through 9 and tap them in the right order, even though the numbers appear randomly on the screen and are replaced by white squares as soon as he starts tapping. Having memorized the numbers, Ayumu touches the squares in the correct order. Reducing the amount of time the numbers flash on the screen doesn't seem to matter to Ayumu, even though humans become less accurate the shorter the time interval. Trying the task myself, I was unable to keep track of more than five numbers after staring at the screen for many seconds, while Ayumu can do the same after seeing the numbers for just 210 milliseconds. This is one-fifth of a

Ayumu's photographic memory allows him to quickly tap a series of numbers on a touchscreen in the right order, even though the numbers disappear in the blink of an eye. That humans cannot keep up with this young ape has upset some psychologists.

second, literally the bat of an eye. One follow-up study managed to train humans up to Ayumu's level with five numbers, but the ape remembers up to nine with 80 percent accuracy, something no human has managed so far.[1] Taking on a British memory champion known for his ability to memorize an entire stack of cards, Ayumu emerged the "chimpion."

The distress Ayumu's photographic memory caused in the scientific community was of the same order as when, half a century ago, DNA studies revealed that humans barely differ enough from bonobos and chimpanzees to deserve their own genus. It is only for historical reasons that taxonomists have let us keep the Homo genus all to ourselves. The DNA comparison caused hand-wringing in anthropology departments, where until then skulls and bones had ruled supremely as the gauge of relatedness. To determine what is important in a skeleton takes judgment, though, which allows the subjective coloring of traits that we deem crucial. We make a big deal of our bipedal locomotion, for example, while ignoring the many animals, from chickens to hopping kangaroos, that move the same way. At some savanna sites, bonobos walk entire distances upright through tall grass, making confident strides like humans.[2] Bipedalism is really not as special as it has been made out to be. The good thing about DNA is that it is immune to prejudice, making it a more objective measure.

With regard to Ayumu, however, it was the turn of psychology departments to be upset. Since Ayumu is now training on a much larger set of numbers, and his photographic memory is being tried on ever shorter time intervals, the limits of what he can do are as yet unknown. But this ape has already violated the dictum that, without exception, tests of intelligence ought to confirm human superiority. As expressed by David Premack, "Humans command all cognitive abilities, and all of them are domain general, whereas animals, by contrast, command very few abilities, and all of them are adaptations restricted to a single goal or activity."[3] Humans, in other words, are a singular bright light in the dark intellectual firmament that is the rest of nature. Other species are conveniently swept together as "animals" or "the animal"—not to mention "the brute" or "the nonhuman"—as if there were no point differentiating among them. It is an us-versus-them world. As the American primatologist Marc Hauser, inventor of the term *humaniqueness*, once said: "My guess is that we will eventually come to see that the gap between human and animal cognition, even a chimpanzee, is greater than the gap between a chimp and a beetle."[4]

You read it right: an insect with a brain too small for the naked eye is put on a par with a primate with a central nervous system that, albeit smaller than ours, is identical in every detail. Our brain is almost exactly like an ape's, from its various regions, nerves, and neurotransmitters to its ventricles and blood supply. From an evolutionary perspective, Hauser's statement is mind-boggling. There can be only one outlier in this particular trio of species: the beetle.

Evolution Stops at the Human Head

Given that the discontinuity stance is essentially pre-evolutionary, let me call a spade a spade, and dub it *Neo-Creationism*. Neo-Creationism is not to be confused with Intelligent Design, which is merely old creationism in a new bottle. Neo-Creationism is subtler in that it accepts evolution but only half of it. Its central tenet is that we descend from the apes in body but not in mind. Without saying so explicitly, it assumes that evolution stopped at the human head. This idea remains prevalent in much of the social sciences, philosophy, and the humanities. It views our mind as so original that there is no point comparing it to other minds except to confirm its exceptional status. Why care about what other species can do if there is literally no comparison with what we do? This saltatory view (from *saltus*, or "leap") rests on the conviction that something major must have happened after we split off from the apes: an abrupt change in the last few million years or perhaps even more recently. While this miraculous event remains shrouded in mystery, it is honored with an exclusive term—hominization—mentioned in one breath with words such as *spark*, *gap*, and *chasm*.[5] Obviously, no modern scholar would dare mention a divine spark, let alone special creation, but the religious background of this position is hard to deny.

In biology, the evolution-stops-at-the-head notion is known as Wallace's Problem. Alfred Russel Wallace was a great English naturalist who lived at the same time as Charles Darwin and is considered the coconceiver of evolution by means of natural selection. In fact, this idea is also known as the Darwin-Wallace Theory. Whereas Wallace definitely had no trouble with the notion of evolution, he drew a line at the human mind. He was so impressed by what he called human dignity that he couldn't stomach comparisons with apes. Darwin believed that all traits were utilitarian, being only as good as strictly necessary for survival, but Wallace felt there must be one exception to this rule: the human mind. Why would people who live simple lives need a brain capable of composing symphonies or doing math? "Natural selection," he wrote, "could only have endowed the savage with a brain a little superior to that of an ape, whereas he actually possesses one but very little inferior to that of the average member of our learned societies."[6] During his travels in Southeast Asia, Wallace had gained great respect for nonliterate people, so for him to call them only "very little inferior" was a big step up over the prevailing racist views of his time, according to which their intellect was halfway between that of an ape and Western man. Although he was nonreligious, Wallace attributed humanity's surplus brain power to the "unseen universe of Spirit." Nothing less could account for the human soul. Unsurprisingly, Darwin was deeply disturbed to see his respected colleague invoke the hand of God, in however camouflaged a way. There was absolutely no need for supernatural explanations, he felt. Nevertheless, Wallace's Problem still looms large in academic circles eager to keep the human mind out of the clutches of biology.

I recently attended a lecture by a prominent philosopher who enthralled us with his take on consciousness, until he added, almost like an afterthought, that "obviously" humans possess infinitely more of it than any other species. I scratched my head—a sign of internal conflict in primates—because until then the philosopher had given the impression that he was looking for an evolutionary account. He had mentioned massive interconnectivity in the brain, saying that consciousness arises from the number and complexity of neural connections. I have heard similar accounts from robot experts, who feel that if enough microchips connect within a computer, consciousness is bound to emerge. I am willing to believe it, even though no one seems to know how interconnectivity produces consciousness nor even what consciousness exactly is.

The emphasis on neural connections, however, made me wonder what to do with animals with brains larger than our 1.35-kilogram brain. What about the dolphin's 1.5-kilogram brain, the elephant's 4-kilogram brain, and the sperm whale's 8-kilogram brain? Are these animals perhaps *more* conscious than we are? Or does it depend on the number of neurons? In this regard, the picture is less clear. It was long thought that our brain contained more neurons than any other on the planet, regardless of its size, but we now know that the elephant brain has three times as many neurons—257 billion, to be exact. These neurons are differently distributed, though, with most of the elephant's in its cerebellum. It has also been speculated that the pachyderm brain, being so huge, has many connections between far-flung areas, almost like an extra highway system, which adds complexity.[7] In our own brain, we tend to emphasize the frontal lobes—hailed as the seat of rationality—but according to the latest anatomical reports, they are not truly exceptional. The human brain has been called a "linearly scaled-up primate brain," meaning that no areas are disproportionally large.[8] All in all, the neural differences seem insufficient for human uniqueness to be a foregone conclusion. If we ever find a way of measuring it, consciousness could well turn out to be widespread. But until then some of Darwin's ideas will remain just a tad too dangerous.

This is not to deny that humans are special—in some ways we evidently are—but if this becomes the a priori assumption for every cognitive capacity under the sun, we are leaving the realm of science and entering that of belief. Being a biologist who teaches in a psychology department, I am used to the different ways disciplines approach this issue. In biology, neuroscience, and the medical sciences, continuity is the default assumption. It couldn't be otherwise, because why would anyone study fear in the rat amygdala in order to treat human phobias if not for the premise that all mammalian brains are similar? Continuity across life-forms is taken for granted in these disciplines, and however important humans may be, they are a mere speck of dust in the larger picture of nature.

Increasingly, psychology is moving in the same direction, but in other social sciences and the humanities discontinuity remains the typical assumption. I am reminded of this every time I address these audiences. After a lecture that inevitably (even if I don't always mention humans) reveals similarities between us and the other Hominoids, the question invariably arises: "But what then does it mean to be human?" The *but* opening is telling as it sweeps all the similarities aside in order to get to the all-important question of what sets us apart. I usually answer with the iceberg metaphor, according to which there is a vast mass of cognitive, emotional, and behavioral similarities between us and our primate kin. But there is also a tip containing a few dozen differences. The natural sciences try to come to grips with the whole iceberg, whereas the rest of academia is happy to stare at the tip.

In the West, fascination with this tip is old and unending. Our unique traits are invariably judged to be positive, noble even, although it wouldn't be hard to come up with a few unflattering ones as well. We are always looking for the one *big* difference, whether it is opposable thumbs, cooperation, humor, pure altruism, sexual orgasm, language, or the anatomy of the larynx. It started perhaps with the debate between Plato and Diogenes about the most succinct definition of the human species. Plato proposed that humans were the only creatures at once naked and walking on two legs. This definition proved flawed, however, when Diogenes brought a plucked fowl to the lecture room, setting it loose with the words "Here is Plato's man." From then on the definition added "having broad nails."

In 1784 Johann Wolfgang von Goethe triumphantly announced that he had discovered the biological roots of humanity: a tiny piece of bone in the human upper jaw known as the *os intermaxillare*. Though present in other mammals, including apes, the bone had never before been detected in our species and had therefore been labeled "primitive" by anatomists. Its absence in humans had been taken as something we should be proud of. Apart from being a poet, Goethe was a natural scientist, which is why he was delighted to link our species to the rest of nature by showing that we shared this ancient bone. That he did so a century before Darwin reveals how long the idea of evolution had been around.

The same tension between continuity and exceptionalism persists today, with claim after claim about how we differ, followed by the subsequent erosion of these claims.[9] Like the *os intermaxillare*, uniqueness claims typically cycle through four stages: they are repeated over and over, they are challenged by new findings, they hobble toward retirement, and then they are dumped into an ignominious grave. I am always struck by their arbitrary nature. Coming out of nowhere, uniqueness claims draw lots of attention while everyone seems to forget that there was no issue before. For example, in the English language (and quite a few others), behavioral copying is denoted by a verb that refers to our closest relatives, hinting at a time when imitation was no big deal and was considered something we shared with the apes. But when imitation was redefined as cognitively complex, dubbed "true imitation," all of a sudden we became the only ones capable of it. It made for the peculiar consensus that we are the only aping apes. Another example is theory of mind, a concept that in fact derives from primate research. At some point, however, it was redefined in such a manner that it seemed, at least for a while, absent in apes. All these definitions and redefinitions take me back to a character played by Jon Lovitz on *Saturday Night Live*, who conjured unlikely justifications of his own behavior. He kept digging and searching until he believed his own fabricated reasons, exclaiming with a self-satisfied smirk, "Yeah! That's the ticket!"

With regard to technical skills, the same thing happened despite the fact that ancient gravures and paintings commonly depicted apes with a walking cane or some other instrument, most memorably in Carl Linnaeus's *Systema Natura* in 1735. Ape tool use was well known and not the least bit controversial at the time. The artists probably put tools in the apes' hands to make them look more humanlike, hence for exactly the opposite reason anthropologists in the twentieth century elevated tools to a sign of brainpower. From then on, the technology of apes was subjected to scrutiny and doubt, ridicule even, while ours was held up as proof of mental preeminence. It is against this backdrop that the discovery (or rediscovery) of ape tool use in the wild was so shocking. In their attempts to downplay its importance, I have heard anthropologists suggest that perhaps chimpanzees learned how to use tools from humans, as if this would be any more likely than having them develop tools on their own. This proposal obviously goes back to a time when imitation had not yet been declared uniquely human. It is hard to keep all those claims consistent. When Leakey suggested that we must either call chimpanzees human, redefine what it is to be human, or redefine tools, scientists predictably embraced the second option. Redefining man will never go out of fashion, and every new characterization will be greeted with "Yeah! That's the ticket!"

Even more egregious than human chest beating—another primate pattern—is the tendency to disparage other species. Well, not just other species, because there is a long history of the Caucasian male declaring himself genetically superior to everyone else. Ethnic triumphalism is extended outside our species when we make fun of Neanderthals as brutes devoid of sophistication. We now know, however, that Neanderthal brains were slightly larger than ours, that some of their genes were absorbed into our own genome, and that they knew fire, burials, handaxes, musical instruments, and so on. Perhaps our brothers will finally get some respect. When it comes to the apes, however, contempt persists. When in 2013 the BBC website asked *Are You as Stupid as a Chimpanzee?* I was curious to learn how they had pinpointed the level of chimpanzee intelligence. But the website (since removed) merely offered a test of human knowledge about world affairs, which had nothing to do with apes. The apes merely served to draw a contrast with our species. But why focus on apes in this regard rather than, say, grasshoppers or goldfish? The reason is, of course, that everyone is ready to believe that we are smarter than these animals, yet we are not entirely sure about species closer to us. It is out of insecurity that we love the contrast with other Hominoids, as is also reflected in angry book titles such as *Not a Chimp* or *Just Another Ape?*[10]

The same insecurity marked the reaction to Ayumu. People watching his videotaped performance on the Internet either did not believe it, saying it must be a hoax, or had comments such as "I can't believe I am dumber than a chimp!" The whole experiment was taken as so offensive that American scientists felt they had to go into special training to beat the chimp. When Tetsuro Matsuzawa, the Japanese scientist who led the Ayumu project, first heard of this reaction, he put his head in his hands. In her charming behind-the-scenes look at the field of evolutionary cognition, Virginia Morrell recounts Matsuzawa's reaction:

Really, I cannot believe this. With Ayumu, as you saw, we discovered that chimpanzees are better than humans at one type of memory test. It is something a chimpanzee can do immediately, and it is one thing—one thing—that they are better at than humans. I know this has upset people. And now there are researchers who have practiced to become as good as a chimpanzee. I really don't understand this need for us to always be superior in all domains.[11]

Even though the iceberg's tip has been melting for decades, attitudes barely seem to budge. Instead of discussing them any further here or going over the latest uniqueness claims, I will explore a few claims that are now close to retirement. They illustrate the methodology behind intelligence testing, which is crucial to what we find. How do you give a chimp—or an elephant or an octopus or a horse—an IQ test? It may sound like the setup to a joke, but it is actually one of the thorniest questions facing science. Human IQ may be controversial, especially while we are comparing cultural or ethnic groups, but when it comes to distinct species, the problems are a magnitude greater.

I am willing to believe a recent study that found cat lovers to be more intelligent than dog lovers, but this comparison is a piece of cake relative to one drawing a contrast between actual cats and dogs. Both species are so different that it would be hard to design an intelligence test that both of them perceive and approach similarly. At issue, however, is not just how two animal species compare but—the big gorilla in the room—how they compare to us. And in this regard, we often abandon all scrutiny. Just as science is critical of any new finding in animal cognition, it is often equally uncritical with regard to claims about our own intelligence. It swallows them hook, line, and sinker, especially if they—unlike Ayumu's feat—are in the expected direction. In the meantime, the general public gets confused, because inevitably any such claims provoke studies that challenge them. Variation in outcome is often a matter of methodology, which may sound boring but goes to the heart of the question of whether we are smart enough to know how smart animals are.

Methodology is all we have as scientists, so we pay close attention to it. When our capuchin monkeys underperformed on a face-recognition task on a touchscreen, we kept staring at the data until we discovered that it was always on a particular day of the week that the monkeys fared so poorly. It turned out that one of our student volunteers, who carefully followed the script during testing, had a distracting presence. This student was fidgety and nervous, always changing her body postures or adjusting her hair, which apparently made the monkeys nervous, too. Performance improved dramatically once we removed this young woman from the project. Or take the recent finding that male but not female experimenters induce so much stress in mice that it affects their responses. Placing a T-shirt worn by a man in the room has the same effect, suggesting that olfaction is key.[12] This means, of course, that mouse studies conducted by men may have different outcomes than those conducted by women. Methodological details matter much more than we tend to admit, which is particularly relevant when we compare species.

Knowing What Others Know

Imagine that aliens from a distant galaxy landed on earth wondering if there was one species unlike the rest. I am not convinced they would settle on us, but let's assume they did. Do you think they'd do so based on the fact that we know what others know? Of all the skills that we possess and all the technology that we have invented, would they zoom in on the way we perceive one another? What an odd and capricious choice this would be! But it is precisely the trait that the scientific community has considered most worthy of attention for the last two decades. Known as *theory of mind*, abbreviated ToM, it is the capacity to grasp the mental states of others. And the profound irony is that our fascination with ToM did not even start with our species. Emil Menzel was the first to ponder what one individual knows about what others know, but he did so for juvenile chimpanzees.

In the late 1960s Menzel would take a young ape by the hand out into a large, grassy enclosure in Louisiana to show her hidden food or a scary object, such as a toy snake. After this, he would bring her back to the waiting group and release them all together. Would the others pick up on the knowledge of one among them, and if so, how would they react? Could they tell the difference between the other having seen food or a snake? They most certainly could, being eager to follow a chimp who knew a food location or being reluctant to stay with one who'd just seen a hidden snake. Copying the other's enthusiasm or alarm, they had an inkling of his knowledge.[13]

Scenes around food were especially telling. If the "knower" ranked below the "guessers," the former had every reason to conceal his or her information to keep the food out of the wrong hands. We recently repeated these experiments with our own chimps and found the same subterfuge as reported by Menzel. Katie Hall would remove two of our chimps from their outdoor enclosure and keep them temporarily in a building. Low-ranking Reinette would have a small window from which to look out into the enclosure, whereas high-ranking Georgia would have no such view. Katie would walk around hiding two food items: one entire banana and one entire cucumber. Guess which one chimps prefer! She'd stuff food underneath a rubber tire, in a hole in the ground, in the deep grass, behind a climbing pole, or some other place, while Reinette followed her every move from inside. Then we'd release both chimps at the same time. By then, Georgia had learned that we'd hide food, but she'd have no clue about the location. She had learned to carefully watch Reinette, who would walk around as nonchalantly as possible while gradually bringing Georgia closer and closer to the concealed cucumber. With Reinette sitting nearby, Georgia would eagerly dig up the veggie. While she was busy, Reinette would hurry toward the banana.

The more experiments we conducted, though, the more Georgia caught on to these deceptive tactics. It is an unwritten rule among chimps that once something is in your hands or mouth, it is yours, even if you are of low status. Before this moment, however, when two individuals approach food, the dominant will enjoy priority. For Georgia, therefore, the trick was to arrive at the banana before Reinette could put her hands on it. After many tests with different combinations of individuals, Katie concluded that high-status chimps exploit the other's knowledge by carefully monitoring their gaze direction, looking where they are looking. Their partners, on the other hand, do their utmost to conceal their knowledge by not looking where they don't want the other to go. Both chimps seem exquisitely aware that one possesses knowledge that the other lacks.[14]

This cat-and-mouse setup shows how much bodies matter. Much of our knowledge about ourselves comes from inside our bodies, and much of what we know about others comes from reading their body language. We are very attuned to the postures, gestures, and facial expressions of others, as are many other animals, such as our pets. This is why Menzel never liked the "theory" language that took over once ToM exploded as a topic as a result of other ape research. The central question became whether apes or children hold a theory about the minds of others.[15] I have trouble with this terminology, too, because it makes it sound as if we understand others through a rational evaluation not unlike the way we figure out physical processes, such as how water freezes or how continents drift apart. It sounds far too cerebral and disembodied. I seriously doubt that we, or any other animal, grasp the mental states of someone else at such an abstract level.

Some even speak of *mindreading*, a term reminiscent of the telepathic trickery of magicians ("Let me guess what card you have in mind"). The magician, however, operates entirely on the basis of which card he has seen you lay your eyes on, or some other visual cue, because there is no such thing as mindreading. All we can do is figure out what others have seen, heard, or smelled, and deduce from their behavior what their next step may be. Putting all this information together is no minor feat and takes extensive experience, but it is body reading, not mindreading. It allows us to look at a situation from the viewpoint of another, which is why I prefer the term *perspective taking*. We use this capacity to our own advantage but also to the advantage of others, such as when we respond to someone else's distress or fulfill the needs of another person. This obviously gets us closer to empathy than ToM.

Human empathy is a critically important capacity, one that holds entire societies together and connects us with those whom we love and care about. It is far more fundamental to survival, I'd say, than knowing what others know. But since it belongs to the large submerged part of the iceberg—traits that we share with all mammals—it doesn't garner the same respect. Moreover, empathy sounds emotional, something cognitive science tends to look down upon. Never mind that knowing what others want or need, or how best to please or assist them, is likely the original perspective taking, the kind from which all other kinds derive. It is essential for reproduction, since mammalian mothers need to be sensitive to the emotional states of their offspring, when they are cold, hungry, or in danger. Empathy is a biological imperative.[16]

Empathic perspective taking, defined by the father of economics, Adam Smith, as "changing places in fancy with the sufferer," is well known outside of our species, including dramatic cases of apes, elephants, or dolphins helping one another under dire circumstances.[17]

Consider how an alpha male chimpanzee at a Swedish zoo saved the life of a juvenile. The juvenile had entangled himself in a rope and was choking to death. The male lifted him up (thus removing the rope's pressure) and carefully unwrapped the rope from his neck. He thus demonstrated an understanding of the suffocating effect of ropes and knew what to do about it. Had he pulled at the juvenile or the rope, he only would have made things worse.

I speak of *targeted helping*, which is assistance based on an appreciation of the other's precise circumstances. One of the oldest reports in the scientific literature concerns an incident, in 1954, off the coast of Florida. During a capture expedition for a public aquarium, a stick of dynamite was set off under the water surface near a pod of bottlenose dolphins. As soon as one stunned victim surfaced, heavily listing, two other dolphins came to its aid: "One came up from below on each side, and placing the upper lateral part of their heads approximately beneath the pectoral fins of the injured one, they buoyed it to the surface in an apparent effort to allow it to breathe while it remained partially stunned." The two helpers were submerged, which meant that they couldn't breathe during the entire effort. The pod remained nearby and waited until

Two dolphins support a third by taking her between them. They buoy the stunned victim so that her blowhole is above the surface, whereas their own blowholes are submerged. After Siebenaler and Caldwell (1956).

their companion recovered, after which they all fled in a hurry, taking tremendous leaps.[18]

Another case of targeted helping occurred one day at Burgers' Zoo. After having cleaned the indoor hall and before releasing the chimps, the keepers hosed out all the rubber tires and hung them one by one on a horizontal log extending from the climbing frame. Upon seeing the tires, female Krom wanted one in which some water remained. The chimps often use tires as vessels to drink from. Unfortunately, this particular tire was at the end of the row, with multiple heavy tires hanging in front of it. Krom pulled and pulled at the one she wanted but was unable to move it. She worked in vain on this problem for over ten minutes, ignored by everyone except Jakie, a seven-year-old that she had taken care of as a juvenile. As soon as Krom gave up and walked away, Jakie approached the scene. Without hesitation he pushed the tires off the log one by one, beginning with the front one, followed by the second in the row, and so on, as any sensible chimp would. When he reached the last tire, he carefully removed it so that no water was spilled and carried it straight to his aunt, placing it upright in front of her. Krom accepted his present without any special acknowledgment and was already scooping up water with her hand when Jakie left.[19]

Having gone over numerous incidents of insightful assistance in *The Age of Empathy*, I am pleased that there are now finally controlled experiments.[20] For example, at the PRI where Ayumu lives, two chimps were placed side by side while one had to guess what kind of tool the other needed to reach attractive food. The first chimp had a choice between a range of tools—such as a straw to suck up juice or a rake to move food closer—only one of which would work for her partner. She'd need to look at and judge her partner's situation before handing her the most useful tool through a window. This is indeed what the chimps did, showing a capacity to grasp the specific needs of others.[21]

The next question is, do primates recognize one another's internal states, such as the difference between a partner who is hungry and one who is sated? Would you give up precious food for someone who has just eaten a big meal right in front of you? This is the question Japanese primatologist Yuko Hattori asked the monkeys in our capuchin colony.

Capuchins can be quite generous and are great social eaters, often sitting in clusters munching together. When a pregnant female hesitates to descend to the floor to collect her own fruits (being arboreal, these monkeys feel safer higher up), we have seen other monkeys grab more than they need and bring handfuls of food up to her. In the experiment, we separated two monkeys with mesh wide enough to stick their arms through, while one of them received a small bucket with apple slices. Under these circumstances, the provisioned monkey often brings food to its empty-handed partner. They sit next to the mesh partition and let the other one reach through to take food out of their hands or mouth, sometimes actively pushing it in their direction. This is remarkable, because the circumstances allow the possessor to avoid sharing altogether by staying away from the mesh. We found one exception to their generosity, however: if their partner had just eaten, the monkeys became stingy. Of course, this could be due to a sated partner being less interested in food, but the monkeys were stingy only if they had actually *seen* their partner eat. A partner that had been fed out of sight was treated as generously as any other. Yuko concluded that the monkeys judged the need, or lack thereof, of their companions based on what they had seen them eat.[22]

In children, an understanding of needs and desires develops years before they realize what others know. They read "hearts" well before they read minds. This suggests that we are on the wrong track in phrasing all this in terms of abstract thinking and theories about others. At a young age, children recognize, for example, that a child looking for his rabbit will be happy to find it, whereas a child searching for his dog will be indifferent to the rabbit.[23] They have an understanding of what others want. Not all humans take advantage of this capacity, which is why we have two kinds of gift-givers: those who go out of their way to find a gift that *you* might like, and those who arrive with what *they* like. Even birds do better than that. In one of those cognitive ripples typical of our field, empathic perspective taking has been suggested for corvids. Male Eurasian jays court their mates by feeding them delicious tidbits. On the assumption that every male likes to impress, experimenters gave him two foods to choose from: wax moth larvae and mealworms. But before giving the male a chance to feed his mate, they would feed her first with one of those two foods. Seeing this, the male would change his choice. If his mate had just eaten a lot of wax moth larvae, he'd pick mealworms for her instead, and vice versa. He did so, however, only if he had witnessed her being fed by the experimenter. Male birds thus took into account what their mate had just eaten, perhaps assuming that she'd be ready for a change of taste.[24] Jays, too, may attribute preferences to others, taking another's point of view.

At this point, you may wonder why perspective taking was ever declared uniquely human. For this, we need to look at a series of ingenious experiments in the 1990s in which chimpanzees could gain information about concealed food either from an experimenter who had witnessed the hiding process or from another one who had been put in the corner with a bucket over his head. Obviously, they should ignore the second experimenter, who had no idea, and follow the directions of the first. They made no distinction, however. Or an ape could beg for cookies from an experimenter sitting out of reach with a blindfold over his eyes. Would the chimps understand that there was no point stretching out an open hand to someone who cannot see them? After a great variety of such tests, the conclusion was that chimpanzees fail to understand the knowledge that others have and don't even realize that knowing requires seeing. It was a most peculiar conclusion, given that the main researcher himself relates how playful apes put buckets or blankets over their heads and walk around until they bump into each other. When he himself put things over his head, however, he immediately became the target of play attacks by these apes, who exploited his obscured vision.[25] They knew he couldn't see them and tried to catch him by surprise.

I knew a couple of juvenile male chimps who loved to throw rocks at us, practicing their impressive long-range aim, invariably doing so as soon as I moved my camera to my eye, which made me lose visual contact. Such behavior alone tells us that apes know something about the vision of others and that tests with blindfolds must therefore be missing something. But as happens so often among experimentalists, behavior in the testing room was given priority over real-life observations. As a result, human exceptionalism was loudly proclaimed, most dramatically by concluding that apes do not possess "anything remotely resembling a ToM."[26]

This conclusion was warmly greeted and is still being broadcast today even though it has not held up to scrutiny. At my home institution, the Yerkes Primate Center, David Leavens and Bill Hopkins conducted tests in which they placed a banana outside a chimpanzee enclosure where humans regularly walked by. Would the chimps draw attention to get people to hand them the fruit? Would they distinguish between people who could see them and those who could not? If so, this would suggest that they grasped another individual's visual perspective. The chimps did, because they'd give visual signals to people who looked in their direction, but they'd vocalize and bang on metal if people failed to notice them. They even pointed at the banana to clarify their wishes. One chimpanzee, afraid to be misunderstood, pointed first with her hand at the banana and then with a finger at her own mouth.[27]

Intentional signaling is not limited to captive apes, as became clear when scientists put a fake snake on the path of wild chimpanzees. Recording the apes' alarm calls in a Ugandan forest, they found that calling is not just a reflection of fear, because the chimps vocalize regardless of whether the snake is near or far. It is rather a warning intended for others: they call more when others are present, especially friends who have failed to notice the serpent. Callers look back and forth between nearby chimps and the danger, calling more to companions who are naïve about it than to those who already know. Callers thus specifically inform those who lack knowledge, likely because they realize how knowing requires seeing.[28]

A critical test of this connection was conducted by Brian Hare, then a student here at the Yerkes Primate Center. Brian wanted to know if apes exploit information about another's visual input. A low-ranking individual was enticed to pick up food in front of a high-ranking one. This is a tricky thing to do, and most subordinates shy away from the confrontation. They were offered a choice between pieces of food that the dominant individual had seen being hidden and pieces hidden without him knowing. The subordinate, on the other hand, had watched it all. In an open competition, like an Easter egg hunt, the safest bet for the subordinate would be to pick only those food items that the dominant had no clue about. This is exactly what they did, showing that they understood that if the dominant had not seen the hiding process, he couldn't know.[29] Brian's study threw the question of animal ToM wide open again. In an unexpected twist, one capuchin monkey at the University of Kyoto and several macaques at a Dutch research center recently passed similar tasks.[30] This is why the whole notion that visual perspective taking is limited to our species is now in the trash bin. Each of the above experiments in and of itself may not be entirely watertight, but taken together they come down on the side of perspective-taking abilities in other species.

It is a testimony to Menzel's pioneering work that we keep hiding food or snakes, and pitting guessers against knowers. It remains the classical paradigm to assess these capacities both in humans and in other species. Perhaps most telling is an experiment by Menzel's son Charles. Like his father, Charlie Menzel is a deep thinker, unsatisfied with easy tests or simple answers. At the Language Research Center here in Atlanta, he'd let a female chimp named Panzee watch while he hid food in the pine forest around her outdoor enclosure. Charlie would dig a small hole in the ground to put a bag of M&Ms into it, or place a candy bar in the bushes. Panzee would follow the process from behind bars. Since she could not go where Charlie was, she would need human help to eventually get the hidden food. Sometimes Charlie would hide it after all other people had gone for the day. This meant Panzee could not communicate with anybody about what she knew until the next morning. When the caretakers arrived, they were unaware of the experiment. Panzee first had to get their attention, then provide information to someone who had no clue as to what she was "talking" about.

During a live demonstration of Panzee's skills, Charlie told me that caretakers generally have a higher opinion of apes' mental abilities than does the typical philosopher or psychologist. This high opinion was essential for his experiment, he explained, because it meant that Panzee was dealing with people who took her seriously. All those recruited by Panzee said they were at first surprised by her behavior but soon understood what she wanted them to do. By following her pointing, beckoning, panting, and calling, they had no trouble finding the candies hidden in the forest. Without her instructions, they would never have known where to look. Panzee never pointed in the wrong direction, or to locations that had been used on previous occasions. The result was communication about a past event, present in the ape's memory, to ignorant members of a different species. If the human followed the instructions correctly and got closer to the food, Panzee would vigorously bob her head in affirmation (like "Yes! Yes!"), and like us, she'd lift her hand up, giving higher points, if the item was farther away. She realized that she knew something that the other didn't know, and was intelligent enough to recruit humans as willing slaves to obtain the goodies of her desire.[31]

Just to illustrate how creative chimps can be in this regard, here is a typical incident at our field station. A young female grunted at me from behind a fence and kept looking at me with shiny eyes (indicating that she knew something exciting) alternated with pointed stares into the grass near my feet. I couldn't figure out what she wanted, until she spat. From the trajectory, I noticed a small green grape. When I gave it to her, she ran to another spot and repeated her performance. Having memorized the locations of fruits dropped by the caretakers, she proved an accurate spitter, collecting three rewards this way.

Clever Hans in Reverse

So why did we at first reach the wrong conclusion about animal perspective taking, and why has it happened so many times before and since? Claims about absent capacities range from the idea that primates do not care about the welfare of others, do not imitate, or even fail to understand gravity. Imagine this for flightless animals that travel high above the ground! In my own career, I have faced resistance to the notion that primates reconcile after fights or console those who are distressed. Or at least I heard the counterclaim that they do not *truly* do so—as in, they do not "truly imitate" or "truly console"—which immediately gets one into debates about how to distinguish what looks like consolation or imitation from the real thing. At times, the overwhelming negativity got to me, as an entire literature burgeoned that was more excited about the cognitive deficits of other species than about their actual accomplishments.[32] It would be like having a career adviser who all the time tells you that you are too dumb for this or too dumb for that. What a depressing attitude!

The fundamental problem with all these denials is that it is impossible to prove a negative. This is no minor issue. When anyone claims the absence of a given capacity in other species, and speculates that it must therefore have arisen recently in our lineage, we hardly need to inspect the data to appreciate the shakiness of such a claim. All we can ever conclude with some certainty is that we have failed to find a given skill in the species that we have examined. We cannot go much further than this, and we certainly may not turn it into an affirmation of absence.

Scientists do so all the time, though, whenever the human-animal comparison is at stake. The zeal to find out what sets us apart overrides all reasonable caution.

Not even with regard to the Monster of Loch Ness or the Abominable Snowman will you ever hear anyone claim to have proven its nonexistence, even though this would fit the expectations of most of us. And why do governments still spend billions of dollars to search for extraterrestrial civilizations while there is no shred of evidence to encourage this quest? Isn't it time to conclude once and for all that these civilizations simply don't exist? But this conclusion will never be reached. That respected psychologists ignore the recommendation to tread lightly around absent evidence is most puzzling, therefore. One reason is that they test apes and children in the same manner—at least in their minds—while coming up with contrary results. Applying a battery of cognitive tasks to both apes and children and finding not a single result in the apes' favor, they tout the differences as proof of human uniqueness. Otherwise, why didn't the apes fare better? To understand the flaw in this logic, we need to go back to Clever Hans, the counting horse. But instead of using Hans to illustrate why animal capacities are sometimes overrated, this time we are concerned with the unfair advantage that human capacities enjoy.

The outcome of ape-child comparisons themselves suggests the answer. When tested on physical tasks, such as memory, causality, and the use of tools, apes perform at about the same level as two-and-a-half-year-old children, but when it comes to social skills, such as learning from others or following others' signals, they are left in the dust.[33] Social problem solving requires interaction with an experimenter, however, whereas physical problem solving does not. This raises the possibility that the human interface is key. The typical format of an experiment is to let apes interact with a white-coated barely familiar human. Since experimenters are supposed to be bland and neutral, they do not engage in schmoozing, petting, or other niceties. This doesn't help make the ape feel at ease and identify with the experimenter. Children, however, are encouraged to do so. Moreover, only the children are interacting with a member of their own species, which helps them even more. Nevertheless, experimenters comparing apes and children insist that all their subjects are treated exactly the same. The inherent bias of this arrangement has become harder to ignore, however, now that we know more about ape attitudes. A recent eye-tracking study (which precisely measured where subjects looked) reached the unsurprising conclusion that apes consider members of their own species special: they follow the gaze of another ape more closely than they follow the human gaze.[34] This may be all we need to explain why apes fare poorly on social tasks presented by members of our species.

There are only a dozen institutes that test ape cognition, and I have visited most of them. I have noticed procedures in which humans barely interact with their subjects and ones in which they have close physical contact. The latter can safely be done only by those who raised the apes themselves or have at least known them since infancy. Since apes are much stronger than we are and have been known to kill people, the up-close-and-personal approach is not for everyone. The other extreme derives from the traditional approach in the psychology lab: carrying a rat or a pigeon into a testing room with as little contact as possible. The ideal here is a nonexistent experimenter, meaning the absence of any personal relation. In some labs apes are called into a room and given only a few minutes to perform before they are sent out again without any playful or friendly contact, almost like a military drill. Imagine if children were tested under such circumstances: how would they fare?

At our center in Atlanta, all our chimps are reared by their own kind and so are more ape- than human-oriented. They are "chimpy," as we say, relative to apes that have a less social background or were raised by humans. We never share the same space with them, but we do interact through the bars, and we always play or groom before testing. We talk to them to put them at ease, give them goodies, and in general try to create a relaxed atmosphere. We want them to look upon our tasks as a game rather than as work, and certainly never put them under pressure. If they are tense because of events in their group, or because another chimp is banging on the outside door or hooting his lungs out, we wait until everyone has calmed down, or we reschedule the test. There is no point testing apes who are not ready. If such procedures are not followed, apes may act as if they don't understand the problem at hand, whereas the real issue is high anxiety and distraction. Many negative results in the literature may be explained this way.

The methodology sections of scientific papers rarely offer a look in the "kitchen," but I think it is crucial. My own approach has always been to be firm and friendly. Firm, meaning that we are consistent and don't make capricious demands but also don't let the animals walk all over us, such as when they only want to play around and get free sweets. But we are also friendly, without punishment, anger, or attempts to dominate. The latter still happens all too often in experiments and is counterproductive with such headstrong animals. Why would an ape follow the points and prompts of a human experimenter whom he sees as a rival? This is another potential source of negative outcomes.

My own team typically cajoles, bribes, and sweet-talks its primate partners. Sometimes I feel like a motivational speaker, such as when Peony, one of our oldest females, ignored a task that we had set up for her. For twenty minutes, she lay in the corner. I sat down right next to her and told her, in a calm voice, that I didn't have all day and it would be great if she would get going. She slowly got up, glancing at me, and strolled to the next room, where she sat down for the task. Of course— as discussed in the previous chapter in relation to Robert Yerkes—it is unlikely that Peony followed the details of what I had said. She was sensitive to my tone of voice and knew all along what we wanted.

However good our relations with apes, the idea that we can test them in exactly the same way we test children is an illusion of the same order as someone throwing both fish and cats into a swimming pool and believing he is treating them the same way. Think of the children as the fish. While testing them, psychologists smile and talk all the time, giving instructions where to look or what to do. "Look at the little froggy!" tells a child so much more than an ape will ever know about the green plastic blob in your hand. Moreover, children are usually tested with a parent in the room, often sitting on their lap. Having permission to run around and an experimenter of their own species, they have an enormous leg up over the ape sitting behind bars without verbal hints or parental support.

True, developmental psychologists try to reduce the influence of parents by telling them not to talk or point, and they may give them sunglasses or a baseball cap to cover their eyes. These measures, however, reveal their woeful underestimation of the power of a parent's motivation to see their child succeed. When it comes to their precious offspring, few people care about the objective truth. We can be glad that Oskar Pfungst designed far more rigorous controls while examining Clever Hans. In fact, Pfungst found that the wide-rimmed hat of the horse owner greatly benefited Hans, since hats amplify head movements. In the same way that the owner vociferously denied his effect on the horse even after it had been proven, parents of children may be completely honest about suppressing cues. But adults have far too many ways to unintentionally guide the choices of a child on their lap, through slight body movements, gaze direction, halted breathing, sighs, squeezes, strokes, and whispered encouragements. Letting parents attend the testing of a child is asking for trouble—the sort of trouble we avoid in animal testing.

The American primatologist Allan Gardner—who was first to teach American Sign Language to an ape—discussed human biases under the heading "Pygmalion leading." Pygmalion, in ancient mythology, was a Cypriot sculptor who fell in love with his own statue of a woman. The story has been used as a metaphor of how teachers raise the performance of certain children by expecting the world of them. They fall in love with their own prediction, which serves as a self-fulfilling prophecy. Remember how Charlie Menzel felt that only people who hold apes in high esteem will fully appreciate what they are trying to communi-

The cognition of children and apes is tested in superficially similar ways. Yet children are not kept behind a barrier; they are talked to and often sit on their parents' laps, all of which helps them connect with the experimenter and receive unintentional hints. The greatest difference, however, is that only apes face a member of another species. Given how much these comparisons disadvantage one class of subjects, they remain inconclusive.

cate? His was a plea for raised expectations, which unfortunately is not the situation apes typically face. Children, in contrast, are treated in such a nurturing manner that they inevitably confirm the mental superiority ascribed to them.[35] Experimenters admire and stimulate them from the outset, making them feel like fish in the water, whereas they often treat apes more like albino rats: keeping them at a distance, and in the dark, while depriving them of the verbal encouragement we offer members of our own species.

Needless to say, I view most ape-child comparisons as fatally flawed.[36]

Recall that apes have been tested for ToM by having them guess what humans know or don't know. The problem here is that captive apes have every reason to believe that we are omniscient! Suppose my assistant calls to tell me that Socko, the alpha male, has been wounded in a fight. I head over to the field station, walk up to him, and ask him to turn around, which he does—having known me since he was a baby—to show me his behind with the gash. Now try to look at this from Socko's perspective. Chimps are smart animals, always trying to figure out what's going on. Of course, he wonders how I know about his injury—I must be an all-knowing god. As such, human experimenters are about the last to be used to find out if apes understand the connection between seeing and knowing. All we are testing is the ape's theory of the *human* mind. It is no accident that we made substantial progress only after egg-hunt scenarios pitted apes against other apes.

One area of cognitive research that has been lucky to escape the species barrier is the study of ToM in animals that are so different from us that everyone understands that humans are unsuitable partners. This has been the case with corvids. Since a true animal watcher never takes a break, the British ethologist Nicky Clayton made a major discovery over lunch at the University of California at Davis. While sitting at an outdoor terrace, she saw Western scrub jays fly off with scraps stolen from the tables. They not only cached them but also guarded them against thieves. If another bird saw where they hid their food, it was bound to

A Western scrub jay caches a mealworm while being watched from behind glass by another. As soon as he is alone, the jay will quickly rehide his treasures, as if realizing that the other knows too much.

disappear. Clayton noticed that after their rivals left the scene, many of the jays returned to rebury their treasures. In follow-up research with Nathan Emery in their lab at Cambridge, she let jays cache mealworms either in private or while being watched by another jay. Given a chance, the jays quickly re-cached their worms at a new location—but only if they had been watched. They seemed to understand that the food was safe if no other birds had any information. Moreover, only birds who themselves had pilfered others' food re-cached their own. Following the dictum "It takes a thief to know a thief," the jays seemed to extrapolate from their own criminality to that of others.[37]

Again, we recognize the Menzel-like design of this experiment, which is even more obvious in a study of perspective-taking ravens. The Austrian zoologist Thomas Bugnyar had a low-ranking male who was expert at opening canisters that contained goodies, but this male often lost his prize to a bullying and stealing dominant male. The low-ranking male, however, learned to distract his competitor by enthusiastically opening empty containers and making as if to eat from them. When the dominant bird found out, "he got very angry, and started throwing things around." Bugnyar further found that when ravens approach hidden food, they take into account what other ravens know. If their competitors have the same knowledge, they hurry to get there first. But if the others are ignorant, they take their time.[38]

All in all, animals do plenty of perspective taking, from being aware of what others want to knowing what others know. A few frontiers are left, of course, such as whether they recognize when others have the *wrong* knowledge. In humans, researchers test this issue with the so-called false-belief task. But since these refinements are hard to evaluate without language, we face a dearth of animal data. Still, even if the remaining differences hold up, there is little doubt that the blanket assertion that ToM is uniquely human must be downgraded to a more nuanced, gradualist view.[39] Humans probably possess a fuller understanding of one another, but the contrast with other animals is not stark enough that extraterrestrials would automatically pick ToM as the chief marker that sets us apart.

While this conclusion is based on solid data from repeated experiments, let me add one anecdote that captures the phenomenon in an entirely different way. At the Yerkes Field Station—where apes live in grassy open-air enclosures in the warm Georgia weather—I developed a special bond with an exceptionally bright female chimp named Lolita. One day Lolita had a new baby, and I wanted to get a good look at it. This is hard to do since a newborn ape is really no more than a little dark blob against its mother's dark tummy. I called Lolita out of her grooming huddle, high up in the climbing frame, and pointed at her belly as soon as she sat down in front of me. Looking at me, she took the infant's right hand in her right hand and its left hand in her left hand. It sounds simple, but given that the baby was ventrally clinging to her, she had to cross her arms to do so. The movement resembled that of people crossing their arms when grabbing a T-shirt by its hems in order to take it off. She then slowly lifted the baby into the air while turning it around its axis, unfolding it in front of me. Suspended from its mother's hands, the baby now faced me instead of her. After it made a few grimaces and whimpers—infants hate to lose touch with a warm belly—Lolita quickly tucked it back into her lap.

With this elegant motion, Lolita demonstrated that she realized I would find the front of her newborn more interesting than its back. To take someone else's perspective represents a huge leap in social evolution.

Spreading Habits

Decades ago friends of mine were outraged by a newspaper article that ranked the smartest canine breeds. They happened to own the breed that was dead last on the list: the Afghan hound. Naturally, the top breed was the border collie. My insulted friends argued that the only reason Afghans were considered dim-witted is that they are independent-minded, stubborn, and unwilling to follow orders. The newspaper's list was about obedience, they said, not intelligence. Afghans are perhaps more like cats, which are not beholden to anyone. This is no doubt why some people rate cats as less intelligent than dogs. We know, however, that a cat's lack of response to humans is not due to ignorance. A recent study showed that felines have no trouble recognizing their owner's voice. The deeper problem is that they don't care, prompting the study's authors to add: "the behavioral aspects of cats that cause their owners to become attached to them are still undetermined."[40]

I had to think of this story when dog cognition emerged as a hot topic. Dogs were depicted as smarter than wolves, perhaps even apes, because they paid better attention to human pointing gestures. A human would point at one out of two buckets, and the dog would check that particular bucket out for a reward. Scientists concluded that domestica-tion had given dogs extra intelligence compared to their ancestors. But what does it mean that wolves fail to follow human pointing? With a brain about one-third larger than a dog's, I bet a wolf could outsmart its domesticated counterpart anytime—yet all we go by is how they react to *us*. And who says that the difference in reaction is inborn, a consequence of domestication, and not based on familiarity with the species doing the pointing? It is the old nature-nurture dilemma. The only way to determine how much of a trait is produced by genes and how much by the environment is to hold one of these two constant to see what *difference* the other one makes. It is a complex problem that is never fully resolved. In the dog-wolf comparison, this would mean raising wolves like dogs in a human household. If they still differ, genetics might be at play.

Raising wolf puppies in the home is a hellish job, though, since they are exceptionally energetic and less rule-bound than dog puppies, chewing up everything in sight. When dedicated scientists raised wolves this way, the nurture hypothesis came out the winner. Human-raised wolves followed hand points as well as dogs. A few differences persisted, though, such as that wolves looked less at human faces than dogs and were more self-reliant. When dogs tackle a problem they cannot solve, they look back at their human companion to get encouragement or assistance—something that wolves never do. Wolves keep trying and trying on their own. Domestication may be responsible for this particular difference. Instead of intelligence, though, it seems more a question of temperament and relations with us—those weird bipedal apes that the wolf evolved to fear and the dog was bred to please.[41] Dogs, for example, engage in lots of eye contact with us. They have hijacked the human parental pathways in the brain, making us care about them in almost the same way that we care about our children. Dog owners who stare into their pet's eyes experience a rapid increase in oxytocin—a neuropeptide involved in attachment and bonding. Exchanging gazes full of empathy and trust, we enjoy a special relationship with the dog.[42]

Cognition requires attention and motivation, yet it cannot be reduced to either. As we have seen, the same problem troubles the comparison between apes and children, an issue that popped up again in the controversy around animal culture. Whereas in the nineteenth century, anthropologists were still open to the possibility of culture outside our own species, in the twentieth they began to write culture with a capital *C* while claiming that the trait is what makes us human. Sigmund Freud considered culture and civilization a victory over nature, while the American anthropologist Leslie White, in a book ironically entitled *The Evolution of Culture*, declared: "Man and culture originated simultaneously—this by definition."[43] Naturally, when the first reports of animal culture came along, defined as habits learned from others—from potato-washing macaques and nut-cracking chimpanzees to bubble-net-hunting humpback whales—they faced a wall of hostility. One line of defense against this offensive notion was to focus on the learning mechanism. If it could be shown that human culture relies on distinct mechanisms, so the thinking went, we might be able to claim culture for ourselves. Imitation became the holy grail of this battle.

To this end, the age-old definition of *imitating* as "doing an act from seeing it done" had to be changed to something narrower, something more advanced. The category *true imitation* was born, which requires one individual to intentionally copy another's specific technique to achieve a specific goal.[44] Merely duplicating behavior, such as one songbird learning another's song, was not enough anymore: it had to be done with insight and comprehension. While imitation is common in lots of animals according to the old definition, true imitation is rare. We learned this fact from experiments in which apes and children were prompted to imitate an experimenter. They'd watch a human model open a puzzle box or rake in food with a tool. While the children copied the demonstrated action, the apes failed, hence the conclusion that other species lack imitative capacities and cannot possibly have culture. The comfort this finding brought to some circles greatly puzzled me, because it did not answer any fundamental questions either about animal culture or about human culture. All it did was draw a flimsy line in the sand.

One can see here the interplay between the redefinition of a phenomenon and the quest to know what sets us apart, but also a deeper methodological problem, because whether apes imitate us or not is wholly beside the point. For culture to arise in a species, all that matters is that its members pick up habits from *one another*. There are only two ways to make a fair comparison in this regard (if we disregard the third option of having white-coated apes administer tests to both apes and children). One is to follow the wolf example: raise apes in a human home so that they are as comfortable as children around a human experimenter. The second is the so-called *conspecific approach*, which is to test a species with models of its own kind.

The first solution yielded results right away, because several human-raised apes turned out to be as good at imitating members of our species as were young children.[45] In other words, apes, like children, are born imitators and prefer to copy the species that raised them. Under most circumstances, this will be their own kind, but if reared by another species, they are prepared to imitate that one as well. Using us as models, these apes spontaneously learn to brush their teeth, ride bicycles, light fires, drive golf carts, eat with a knife and fork, peel potatoes, and mop the floor. It reminds me of suggestive stories on the Internet about dogs raised by cats, which show feline behavior such as sitting in boxes, crawling under tight spaces, licking their paws to clean their face, or sitting with their front legs tucked in.

Another critical study was conducted by Victoria Horner, a Scottish primatologist, who later became my team's lead expert on cultural learning. Together with Andrew Whiten of St. Andrews University, Vicky worked with a dozen orphan chimps at Ngamba Island, a sanctuary in Uganda. She acted like a mix between a mother and caretaker for the juvenile apes. Sitting next to her during tests, the juvenile apes were attached to Vicky and eager to follow her example. Her experiment created waves because as in Ayumu's case, the apes proved to be smarter than the children. Vicky would poke a stick into holes in a large plastic box, going through a series of holes until a candy would roll out. Only one hole mattered. If the box was made of black plastic, it was impossible to tell that some of the holes were just for show. A transparent box, on the other hand, made it obvious where the candies came from. Handed the stick and the box, young chimps mimicked only the necessary moves, at least with the transparent box. The children, on the other hand, mimicked everything that Vicky had demonstrated, including useless moves. They did so even with the transparent box, approaching the problem more like a magic ritual than as a goal-directed task.[46]

With this outcome, the whole strategy of redefining imitation backfired! After all, it was the apes who best fit the new definition of true imitation. The apes were showing *selective imitation*, the sort that pays close attention to goals and methods. If imitation requires understanding, we have to give it to the apes, not to the children, who for lack of a better term, showed only dumb copying.

What to do now? Premack complained that it was way too easy to make children look "foolish"—as if that were the goal of the experiment!—whereas in reality, he felt, there must be something wrong with the interpretation.[47] His distress was genuine, showing to what degree the human ego gets in the way of dispassionate science. Promptly, psychologists settled on a narrative in which *overimitation*—a new term for children's indiscriminate copying—is actually a brilliant achievement. It fits our species' purported reliance on culture, because it makes us imitate behavior regardless of what it is good for; we transmit habits in full, without every individual making his or her own ill-informed decisions. Given the superior knowledge of adults, the best strategy for a child is to copy them without question. Blind faith is the only truly rational strategy, it was concluded with some relief.

Even more striking were Vicky's studies at our field station in Atlanta, where we started a decade-long research program in collaboration with Whiten, focusing entirely on the conspecific approach. When chimps were given a chance to watch one another, incredible talents for imitation manifested themselves. Apes truly do ape, allowing behavior to be faithfully transmitted within the group.[48] A video of Katie imitating her mother, Georgia, offers a nice example. Georgia had learned to flip open a little door in a box, then stick a rod deep into the opening to retrieve a reward. Katie had watched her mother do this five times, following her every move and smelling Georgia's mouth every time she got a reward. After her mother was moved to another room, Katie could finally access the box herself. Even before we had added any rewards, she flipped open the door with one hand and inserted the rod with the other. Sitting like this, she looked up at us on the other side of the window and impatiently rapped it, while grunting, as if telling us to hurry up. As soon as we pushed the reward into the box, she retrieved it. Before ever being rewarded for these actions, Katie perfectly duplicated the sequence she had watch Georgia perform.

Rewards are often secondary. Imitation without reward is of course common in human culture, such as when we mimic hairstyles, accents, dance steps, and hand gestures, but it is also common in the rest of the primate order. The macaques on the Arashiyama mountaintop in Japan customarily rub pebbles together. The young learn to do it without any reward other than perhaps the noise associated with it. If one case refutes the common notion that imitation requires reward, it is this weird behavior, about which Michael Huffman, an American primatologist who has studied it for decades, notes, "It is likely that the infant is first exposed *in utero* to the click-clacking sounds of stones as its mother plays, and then exposed visually as one of the first activities it sees after birth, when its eyes begin to focus on objects around it."[49]

The word *fashion* was first used in relation to animals by Köhler, whose apes invented new games all the time. They'd march single file around and around a post, trotting in the same rhythm with emphasis on one stamping foot, while the other foot stepped lightly, wagging their heads in the same rhythm, all acting in synchrony as if in a trance. For months our own chimps had a game we called cooking. They'd dig a hole in the dirt, collect water by holding a bucket under a faucet, and dump water into the hole. They'd sit around the hole poking in the mud with a stick as if stirring soup. Sometimes there were three or four such holes in operation at the same time, keeping half the group busy. At a chimpanzee sanctuary in Zambia, scientists followed the spread of yet another meme. One female was the first to stick a straw of grass into her ear, letting it hang out while walking around and grooming others. Over the years, other chimps followed her example, with several of them adopting the same new "look."[50]

Fashions come and go in chimps as in humans, but some habits we find in only one group and not in another. Typical is the hand-clasp grooming of some wild chimpanzee communities, in which two individuals hold hands above their heads while grooming each other's armpits with their other hands.[51] Since habits and fashions often spread without any associated rewards, social learning is truly social. It is about conformity instead of payoffs. Thus an infant male chimp may mimic the charging display of the alpha male who always bangs a specific metal door to accentuate his performance. Ten minutes after the male has finished his performance—a dangerous activity, during which mothers keep their children near—the little son is let go. With all his hair on end, he goes to bang on the same door as his role model.

Having documented numerous such examples, I developed the idea of *Bonding- and Identification-based Observational Learning* (BIOL). Accordingly, primate social learning stems from an urge to belong. BIOL refers to conformism born from the desire to act like others and to fit in.[52] It explains why apes imitate their own kind far better than the average human, and why, among humans, they imitate only those whom they feel close to. It also explains why young chimps, especially females,[53] learn so much from their mothers, and why high-status individuals are favorite models. This preference is also known in our own societies, in which advertisements feature celebrities showing off watches, perfumes, and cars. We love to emulate the Beckhams, Kardashians, Biebers, and

Jolies. Might the same apply to apes? In one experiment, Vicky spread brightly colored plastic chips around in an enclosure, which the chimps could collect and carry to a container in exchange for rewards. Exposed to the sight of a top-ranking group member trained to drop tokens into one container and a bottom-ranking one trained to use a different container, the colony massively followed in the footsteps of the more prestigious member.[54]

As evidence mounted regarding imitation in apes, other species inevitably joined the ranks, showing similar capacities.[55] There are now compelling studies on imitation in monkeys, dogs, corvids, parrots, and dolphins. And if we take a broader view, we have even more species to consider because cultural transmission is widespread. To return to dogs and wolves, a recent experiment applied the conspecific approach to canine imitation. Instead of following human instructions, both dogs and wolves saw a member of their own species manipulate a lever to open the lid of a box with hidden food. Next, they were allowed to try the same box themselves. This time the wolves greatly outsmarted the dogs.[56] Wolves may be poor at following *human* pointing, but when it comes to picking up hints from their own kind, they beat dogs. The investigators ascribe this contrast to attention rather than cognition. They point out that wolves watch one another more closely as they rely on the pack for survival, whereas dogs rely on us.

Clearly, it is time for us to start testing animals in accordance with their biology and move away from human-centric approaches. Instead of making the experimenter the chief model or partner, we better keep him or her in the background. Only by testing apes with apes, wolves with wolves, and children with human adults can we evaluate social cognition in its original evolutionary context. The one exception may be the dog, which we domesticated (or which domesticated itself, as some believe) to bond with us. Humans testing dog cognition may actually be a natural thing to do.

Moratorium

Having escaped the Dark Ages in which animals were mere stimulus-response machines, we are free to contemplate their mental lives. It is a great leap forward, the one that Griffin fought for. But now that animal cognition is an increasingly popular topic, we are still facing the mindset that animal cognition can be only a poor substitute of what we humans have. It can't be truly deep and amazing. Toward the end of a long career, many a scholar cannot resist shining a light on human talents by listing all the things we are capable of and animals not.[57] From the human perspective, these conjectures may make a satisfactory read, but for anyone interested, as I am, in the full spectrum of cognitions on our planet, they come across as a colossal waste of time. What a bizarre animal we are that the only question we can ask in relation to our place in nature is "Mirror, mirror on the wall, who is the smartest of them all?"

Keeping humans in their preferred spot on that absurd scale of the ancient Greeks has led to an obsession with semantics, definitions, and redefinitions, and—let's face it—the moving of goalposts. Every time we translate low expectations about animals into an experiment, the mirror's favorite answer sounds. Biased comparisons are one ground for suspicion, but the other is the touting of absent evidence. I have lots of negative findings in my own drawers that have never seen the light since I have no idea what they mean. They may indicate the absence of a given capacity in my animals, but most of the time, especially if spontaneous behavior suggests otherwise, I am unsure that I have tested them in the best possible way. I may have created a situation that threw them off, or presented the problem in such an incomprehensible fashion that they didn't even bother to solve it. Recall the low opinion scientists held of gibbon intelligence before their hand anatomy was taken into account, or the premature denials of mirror self-recognition in elephants based on their reaction to an undersize mirror. There are so many ways to account for negative outcomes that it is safer to doubt one's methods before doubting one's subjects.

Books and articles commonly state that one of the central issues of evolutionary cognition is to find out what sets us apart. Entire conferences have been organized around the human essence, asking "What makes us human?" But is this truly the most fundamental question of our field? I beg to differ. In and of itself, it seems an intellectual dead end. Why would it be any more critical than knowing what sets cockatoos or beluga whales apart? I am reminded of one of Darwin's random musings: "He who understands baboon would do more towards metaphysics than Locke."[58] Every single species has profound insights to offer, given that its cognition is the product of the same forces that shaped ours. Imagine a medical textbook that declared that its discipline's central issue is to find out what is unique about the human body. We would roll our eyes, because even though this question is mildly intriguing, medicine faces far more basic issues related to the functioning of hearts, livers, cells, neural synapses, hormones, and genes.

Science seeks to understand not the rat liver or the human liver but the liver, period. All organs and processes are a great deal older than our species, having evolved over millions of years with a few modifications specific to each organism. Evolution always works like this. Why would cognition be any different? Our first task is to find out how cognition in general operates, which elements it requires to function, and how these elements are attuned to a species's sensory systems and ecology. We want a unitary theory that covers all the various cognition*s* found in nature. To create space for this project, I recommend placing a moratorium on human uniqueness claims. Given their miserable track record, it is time to rein them in for a few decades. This will allow us to develop a more comprehensive framework. One day years from now, we may then return to our species's particular case armed with new concepts that allow a better picture of what is special—and what not—about the human mind.

One aspect we might focus on during this moratorium is an alternative to overly cerebral approaches. I have already mentioned that perspective taking is likely tied to bodies, and the same applies to imitation. After all, imitation requires that another individual's body movements are perceived and translated into one's own body movements. Mirror neurons (special neurons in the motor cortex that map another's actions onto one's own bodily representations in the brain) are often thought to mediate this process, and it is good to realize that those neurons were discovered not in humans but in macaques. Even though the precise connection remains a point of debate, imitation likely is a bodily process facilitated by social closeness.

This view is quite different from the cerebral one according to which it all depends on the understanding of cause-effect relations and goals. Thanks to an ingenious experiment by the British primatologist Lydia Hopper, we know which view is correct. Hopper presented chimps with a so-called ghost box controlled by fishing lines. The box magically opened and closed by itself, producing rewards. If technical insight were all that mattered, watching such a box should suffice, as it shows all the necessary actions and consequences. But in fact, letting chimps watch the ghost box ad nauseam taught them nothing. Only after seeing an actual chimp operate the same box, did they learn how to get the rewards.[59] Thus for imitation to occur, apes need to connect to a moving body, preferably one of their own species. Technical understanding is not the key.[60]

To find out how bodies interact with cognition, we have incredibly rich material to work with. Adding animals to the mix is bound to stimulate the up-and-coming field of "embodied cognition," which postulates that cognition reflects the body's interactions with the world. Until now, this field has been rather human-focused while failing to take advantage of the fact that the human body is only one of many.

Consider the elephant. It combines a very different body with the brainpower to achieve high cognition. What is the largest land mammal doing with three times as many neurons as our own species? One may downplay this number, arguing that it has to be corrected for body mass, but such corrections are more suited to brain weight than to number of neurons. In fact, it has been proposed that absolute neuron count, regardless of brain or body size, best predicts a species' mental powers.[61] If so, we'd better pay close attention to a species that has vastly more neurons than we do. Since most of these neurons reside in the elephant's cerebellum, some feel they carry less weight, the assumption being that only the prefrontal cortex matters. But why take the way our brain is organized as the measure of all things and look down on subcortical areas?[62] For one thing, we know that during Hominoid evolution, our cerebellum expanded even more than our neocortex. This suggests that for our species, too, the cerebellum is critically important.[63] It is now up to us to find out how the remarkable neuron count of the elephant brain serves its intelligence.

The trunk, or proboscis, is an extraordinarily sensitive smelling, grasping, and feeling organ said to contain forty thousand muscles coordinated by a unique proboscis nerve that runs along its full length. The trunk has two sensitive "fingers" at the tip, with which it can pick up items as small as a blade of grass, but the trunk also allows the animal to suck up eight liters of water or flip over an annoying hippo. True, the cognition associated with this appendage is specialized, but who knows how much of our own cognition is tied to the specifics of our bodies, such as our hands? Would we have evolved the same technical skills and intelligence without these supremely versatile appendages? Some theories of language evolution postulate its origin in manual gestures as well as in neural structures for the throwing of stones and spears.[64] In the same way that humans have a "handy" intelligence, which we share with other primates, elephants may have a "trunky" one.

There is also the issue of continued evolution. It is a widespread misconception that humans kept evolving while our closest relatives stopped. The only one who stopped, however, is the *missing link*: the last common ancestor of humans and apes, so named because it went extinct long ago. This link will forever remain missing unless we happen to dig up some fossil remains. I named my research center Living Links, in a wordplay on the missing link, since we study chimpanzees and bonobos as live links to the past. The name has caught on, because there are now a few other Living Links centers in the world. Traits shared across all three species—our two closest ape relatives and ourselves—likely have the same evolutionary roots.

But apart from commonalities, all three species also evolved in their own separate ways. Since there is no such thing as halted evolution, all three probably changed substantially. Some of these evolutionary changes gave our relatives an advantage, such as the resistance to the HIV-1 virus that evolved in West African chimps long before the AIDS epidemic devastated humanity.[65] Human immunity has some serious catching up to do. Similarly, all three species—not just ours—had time to evolve cognitive specializations. No natural law says that our species has to be best at everything, which is why we should be prepared for more discoveries such as Ayumu's flash memory or the selective imitative talents of apes. A Dutch educational program recently brought out an advertisement in which human children face the floating peanut task (see Chapter 3). Even though the members of our species have a bottle of water standing not too far away, they fail to think of the solution until they see a video of apes solving the same problem. Some apes do so spontaneously, even when there is no bottle around to suggest what to do. They walk to the faucet where they know water can be collected. The point of the ad is that schools should teach kids to think outside the box, using apes as an inspiration.[66]

The more we know about animal cognition, the more examples of this kind may come to light. The American primatologist Chris Martin, at the PRI in Japan, has added yet another chimpanzee forte. Using separate computer screens, he had apes play a competitive game that required them to anticipate one another's moves. Could they outguess their rivals based on their previous choices, a bit like the rock-paper-scissors game? Martin had humans play the same game. The chimps outperformed the humans, reaching optimal performance more quickly and completely than members of our own species. The scientists attributed the edge to chimps being quicker at predicting a rival's moves and countermoves.[67]

This finding resonated with me, given what I know about the politics and preemptive tactics of chimpanzees. Chimp status is based on alliances, in which males support one another. Reigning alpha males protect their power by a divide-and-rule strategy, and they particularly hate it when one of their rivals cozies up to one of their own supporters. They try to forestall hostile collusions. Moreover, not unlike presidential candidates who hold babies up in the air as soon as the cameras are rolling, male chimps vying for power develop a sudden interest in infants, which they hold and tickle in order to curry favor with the females.[68] Female support can make a huge difference in rivalries among males, so making a good impression on them is important. Given the tactical shrewdness of chimpanzees, it is a great advance that computer games now help us put these remarkable skills to the test.

We have no good reason to focus solely on chimpanzees, though. They often serve as a starting point, but "chimpocentrism" is a mere extension of anthropocentrism.[69] Why not focus on other species that lend themselves to explore specific aspects of cognition? We could focus on a small number of organisms as test cases. We already do so in medicine and general biology. Geneticists exploit fruit flies and zebra fish, and students of neural development have gotten much mileage out of research on nematode worms. Not everyone realizes that science works this way, which is why scientists were dumbfounded by the complaint of former vice-presidential candidate Sarah Palin that tax dollars were going to useless projects such as "fruit fly research in Paris, France. I kid you not."[70] It may sound silly to some, but the humble *Drosophila* has long been our main workhorse in genetics, yielding insight in the relation between chromosomes and genes. A small set of animals produces basic knowledge applicable to many other species, including ourselves. The same applies to cognitive research, such as the way rats and pigeons have shaped our view of memory. I imagine a future in which we explore a range of capacities in specific organisms on the assumption of generalizability. We may end up studying technical skills in New Caledonian crows and capuchin monkeys, conformity in guppies, empathy in canids, object categorization in parrots, and so on.

Yet all this requires that we circumvent the fragile human ego and treat cognition like any other biological phenomenon. If cognition's basic features derive from gradual descent with modification, then notions of leaps, bounds, and sparks are out of order. Instead of a gap, we face a gently sloping beach created by the steady pounding of millions of waves. Even if human intellect is higher up on the beach, it was shaped by the same forces battering the same shore.

6 | SOCIAL SKILLS

The old male faced a choice worthy of a politician. Every day Yeroen was being groomed by two rivaling males, each one eager to gain his backing. He seemed to enjoy the attention. Being groomed by the mighty alpha male, the one who had deposed him a year earlier, was utterly relaxing because no one would dare disturb them. But being groomed by the second, younger male was tricky. Their get-togethers greatly upset alpha, who regarded them as plots against himself and tried to disrupt them. Alpha would put up all his hair and hoot and display around, banging doors and hitting females, until the other two males became so nervous that they'd break up and leave the scene. Separating them was the only way to calm down alpha. Since male chimps never cease to jockey for position and are always making and breaking pacts, innocent grooming sessions don't really exist. Every single one carries political implications.

The current alpha male enjoyed massive popularity and support, including that of the old matriarch, Mama, leader of the females. If Yeroen had wanted an easy life, he would have opted to play sidekick to this male. He wouldn't have rocked the boat, and there would never have been any threat to his position. Aligning himself with the ambitious young male, on the other hand, was fraught with risk. However big and muscular this male might be, he had barely left his adolescence behind. He was an untried entity who carried so little authority that whenever he tried to break up a female fight, as top males are wont to do, he risked the wrath of both contestants. Ironically, this meant that he did resolve the discord, but at his own expense. Instead of screaming at one another, the females now supported one another in chasing the would-be arbitrator. Once they got him cornered, however, they were smart enough not to physically grapple with him, being all too familiar with his speed, strength, and canine teeth. He had become a player to be reckoned with.

The alpha male, in contrast, was so skilled at peacekeeping, so impartial in his interventions, and so protective of the underdog that he had become immensely beloved. He had brought peace and harmony to the group after a long period of upheaval. Females were always ready to groom him and let him play with their children. They were likely to resist anyone who dared challenge his reign.

Nonetheless, this is exactly what Yeroen went for when he sided with the young upstart. The two of them entered a long campaign to dethrone the established leader that took a great toll in tensions and injuries. Whenever the young male would position himself at some distance from the alpha male, provoking him with increasingly loud hooting, Yeroen would go sit right behind the challenger, wrap his arms around his middle, and softly hoot along. This way there was no doubt about his allegiance. Mama and her female friends did resist this revolt, occasionally resulting in massive pursuits of both troublemakers, but the combination of the young male's brawn and Yeroen's brain was too much. From the start, it was obvious that Yeroen was not out to claim the alpha position for himself but was content to let his partner do the dirty work. They never backed down, and after several months of daily confrontations, the young male became the new alpha.

The two of them ruled for years, with Yeroen acting like a Dick Cheney or a Ted Kennedy, a power behind the throne; he remained so influential that as soon as his support began to waver, the throne wobbled. This happened occasionally after conflicts over sexually attractive females. The new alpha quickly learned that in order to keep Yeroen on his side, he'd need to grant him privileges. Most of the time Yeroen was allowed to mate with females, something the young alpha did not tolerate from any other male.

Why did Yeroen throw his support behind this parvenu instead of joining the established power? It is informative to look at studies of human coalition formation, in which players win games through cooperation, and to study the balance-of-power theories about international pacts. The basic principle here is the "strength is weakness" paradox, according to which the most powerful player is often the least attractive political ally because this player doesn't really need others, hence takes them for granted and treats them like dirt. In Yeroen's case, the established alpha male was too mighty for his own good. By joining him, Yeroen would have gained little, because all this male truly needed was his neutrality. The smarter strategy was to pick a partner who couldn't win without him. By throwing his weight behind the young male, Yeroen became the kingmaker. He regained both prestige and fresh mating opportunities.

Machiavellian Intelligence

When I began observing the world's largest chimpanzee colony, at Burgers' Zoo in 1975, I had no idea that I'd be working with this species for the rest of my life. Just so, as I sat on a wooden stool watching primates on a forested island for an estimated ten thousand hours, I had no idea that I'd never again enjoy that luxury. Nor did I realize that I would develop an interest in power relations. In those days, university students were firmly antiestablishment, and I had the shoulder-long hair to prove it. We considered ambition ridiculous and power evil. My observations of the chimps, however, made me question the idea that hierarchies were merely cultural institutions, a product of socialization, something we could wipe out at any moment. They seemed more ingrained. I had no trouble detecting the same tendencies in even the most hippielike organizations. They were generally run by young men who mocked authority and preached egalitarianism yet had no qualms about ordering everyone else around and stealing their comrades' girlfriends. It wasn't the chimps who were odd, but the humans who seemed dishonest. Political leaders have a habit of concealing their power motives behind nobler desires such as a readiness to serve the nation and improve the economy. When the English political philosopher Thomas Hobbes postulated the existence of an insuppressible power drive, he was right on target for both humans and apes.

The biological literature proved to be of no help understanding the social maneuvering that I observed, so I turned to Niccolò Machiavelli. During quiet moments of observation, I read from a book that had been published more than four centuries earlier. *The Prince* put me in the right frame of mind to interpret what I was seeing on the chimpanzees' forested island, though I'm pretty sure the Florentine philosopher never envisioned this particular application.

Among chimpanzees, hierarchy permeates everything. Whenever we set out to bring two females inside the building—as we often do for testing—one will be ready to get going on the task at hand while the other will hang back. The second female will barely take rewards and won't touch the puzzle box, computer, or whatever else we're using. She may be just as eager as the other, but she defers to her "superior." There is no tension or hostility between them, and out in the group they may be the best of friends. One female simply dominates the other.

Among the males, in contrast, power is always up for grabs. It is not conferred on the basis of age or any other trait but has to be fought for and jealously guarded against contenders. Soon after my long stint as chronicler of their social affairs, I put pencil to paper to produce *Chimpanzee Politics*, a popular account of the power struggles that I had witnessed.[1] I was risking my nascent academic career by ascribing intelligent social maneuvering to animals, an implication I had been trained to avoid at all cost. That doing well in a group full of rivals, friends, and relatives requires considerable social skill is something we now take for granted, but in those days animal social behavior was rarely thought of as intelligent. Observers would recount a rank reversal between two baboons, for example, in passive terms, as if it happened *to* them rather than was brought about *by* them. They would make no mention of one baboon following the other around, provoking one confrontation after another, flashing his huge canine teeth, and recruiting help from nearby males. It is not that the observers did not notice, but animals were not supposed to have goals and strategies, so the reports remained silent.

Deliberately breaking with this tradition, describing chimps as schmoozing and scheming Machiavellians, my book drew wide attention and enjoyed many translations. The U.S. Speaker of the House, Newt Gingrich, even put it on the recommended reading list for freshmen congressmen. The account met with far less resistance than I had dreaded, including from fellow primatologists. Obviously, the time was ripe, in 1982, for a more cognitive approach to animal social behavior. Even though I learned about it only after my own book, Donald Griffin's *Animal Awareness* had come out just a few years before.[2]

My work was part of a new *Zeitgeist*, and I had a handful of predecessors to lean on. There was Emil Menzel, whose work on chimpanzee cooperation and communication postulated goals and hinted at intelligent solutions, and Hans Kummer, who never ceased to wonder what drove his baboons to act the way they did. Kummer wanted to know, for example, how baboons plan their travel routes, and who decides where to go—those in front or those in the back? He broke down the behavior into recognizable mechanisms, and stressed how social relationships serve as long-term investments. More than anyone before him, Kummer combined classical ethology with questions about social cognition.[3]

I was also impressed by *In The Shadow of Man* by a young British primatologist.[4] By the time I read it, I was familiar enough with chimpanzees to be unsurprised by the specifics of Jane Goodall's description of life at Gombe Stream in Tanzania. But the tone of her account was truly refreshing. She did not necessarily spell out the cognition of her subjects, but it was impossible to read about Mike—a rising male who impressed his rivals by loudly banging empty kerosene cans together— or the love life and family relations of matriarch Flo, without recognizing a complex psychology. Goodall's apes had personalities, emotions, and social agendas. She did not unduly humanize them, but she related what they did in unpretentious prose that would have been perfectly normal for a day at the office but was unorthodox with regard to animals. It was a huge improvement over the tendency at the time to drown behavioral descriptions in quotation marks and dense jargon in order to avoid mentalistic implications. Even animal names and genders were often avoided. (Every individual was an "it.") Goodall's apes, in contrast, were social agents with names and faces. Rather than being the slaves of their instincts, they acted as the architects of their own destinies. Her approach perfectly fit my own budding understanding of chimpanzee social life.

Yeroen's allegiance to the young alpha was a case in point. Not that I could resolve how and why he had made his choice, in the same way that it was impossible for Goodall to know if Mike's career might have been different in the absence of kerosene cans, but both stories implied deliberate tactics. Pinpointing the cognition behind such behavior requires collecting a mass of systematic data as well as performing experiments, such as the strategic computer games that we now know chimps are extraordinarily good at.[5]

Let me briefly offer two examples of how these issues may be tackled. The first concerns a study at the Burgers' Zoo itself. Conflicts in the colony rarely remained restricted to the original two contestants, since chimps have a tendency to draw others into the fray. Sometimes ten or more chimps would be running around, threatening and chasing one another, uttering high-pitched screams that could be heard a mile away. Naturally, every contestant tried to get as many allies on his or her side as possible. When I analyzed hundreds of videotaped incidents (a new technique at the time!), I found that the chimpanzees who were losing the battle beseeched their friends by stretching out an open hand to them. They tried to recruit support in order to turn things around. When it came to the friends of their enemies, however, they went out of their way to appease them by putting an arm around them and kissing their face or shoulder. Instead of begging for assistance, they sought to neutralize them.[6]

To know the friends of your opponents takes experience. It implies that individual A is aware not only of her own relations with B and C but also of the relation between B and C. I dubbed this *triadic awareness*, since it reflects knowledge of the entire ABC triangle. It is the same with us, when we realize who is married to whom, who is a son of whom, or who is the employer of whom. Human society could not function without triadic awareness.[7]

The second example concerns wild chimpanzees. It is well known that there is no obvious connection between a male's rank and his size— the biggest, meanest male does not automatically reach the top. A small male with the right friends also has a shot at the alpha position. This is why male chimps put so much effort into alliance formation. In an analysis of years of data collected at Gombe, a relatively small alpha male spent far more time grooming others than did larger males in the same position. Apparently, the more a male's position depends on support from third parties, the more energy he needs to invest into diplomacy, such as grooming.[8] In a study in the Mahale Mountains, not far from Gombe, Toshisada Nishida and his team of Japanese scientists observed an alpha male with an exceptionally long tenure of more than a decade. This male developed a "bribery" system, selectively sharing prized monkey meat with his loyal allies, while denying such favors to his rivals.[9]

Years after *Chimpanzee Politics*, these studies confirmed the tit-for-tat deal making that I had implied. But even while I was writing my book, supportive data were being gathered. Unknown to me, Nishida had followed an older male at Mahale, named Kalunde, who had moved himself into a key position by playing off younger, competitive males against one another. These young males sought Kalunde's support, which he handed out rather erratically, making himself indispensable to the advancement of any one of them. Being the dethroned alpha male, Kalunde made a comeback of sorts, but like Yeroen, he didn't claim the top position for himself. He rather acted as power behind the scenes. The situation was so eerily similar to the saga I had described that I was thrilled, two decades later, to meet Kalunde in person. Toshi, as the late Nishida was known to his friends, invited me for some fieldwork, which I gladly accepted. He was one of the world's greatest chimpanzee experts, and it was a treat to follow him around through the jungle.

Living in the camp near Lake Tanganyika, one realizes that running water, electricity, toilets, and telephones are greatly overrated. It is entirely possible to survive without them. Every day the goal was to get up early, eat a quick breakfast, and get going before the sun rose. The chimps would have to be found, and the camp had several trackers to assist us. Fortunately, chimps are incredibly noisy, which makes them easy to locate. Chimps do not travel all in a single group but are spread out over separately traveling "parties" of just a few individuals each. In an environment with low visibility, they rely heavily on vocalizations to stay in touch. Following an adult male, for example, you continuously see him stop, cock his head, and listen to others in the distance. You see him decide how to respond, by replying with his own calls, silently moving toward the source (sometimes in such a hurry that you are left struggling through tangled vines), or continue on his merry way as if what he just heard lacked any relevance.

By then Kalunde was the oldest male, only about half the size of a prime adult male. Being around forty, he had shrunk. But despite his advanced age, he was still into political games, frequently accompanying and grooming the beta male until alpha returned from a long period of absence. Alpha had traveled to the fringes of the community territory, escorting a sexually receptive female. High-ranking males may go for weeks on end "on safari" with a female, as it is known, in order to avoid competition. I knew about alpha's unexpected return only because Toshi told me in the evening, but I had noticed great agitation in the males that I had been following the whole day. They were restless, running up and down the hills, totally exhausting me. Alpha's characteristic hooting and drumming on empty trees had announced his return, making everyone hypernervous. In the following days, it was fascinating to see Kalunde switching camps. One moment he would be grooming the returning alpha; the next he'd be hanging out with the beta male, as if trying to decide which side he should be on. He offered the perfect illustration of a tactic that Toshi had dubbed "allegiance fickleness."[10]

You can imagine that we had much to talk about, especially comparing wild versus zoo chimps. Obviously, there are major differences, but it is not as simple as some people think, especially those who wonder why one would study captive animals at all. The goals of both types of research are quite different, and we need both. Fieldwork is essential to understanding the natural social life of any animal. For anyone who wants to know how and why their typical behavior evolved, there is no substitute for observing them in their natural habitat. I have visited many field sites, from capuchin monkeys in Costa Rica and woolly spider monkeys in Brazil to orangutans in Sumatra, baboons in Kenya, and Tibetan macaques in China. I find it very informative to see the ecology of wild primates and to hear from colleagues what sort of issues they are fascinated by. Fieldwork is nowadays very systematic and scientific. The days of a few scribbled observations in a notebook are gone. Data collection is continuous and systematic, typed into handheld digital devices, and complemented with fecal and urine samples that allow DNA analysis and hormone assays. All this hard, sweaty work has enormously advanced our understanding of wild animal societies.

Yet in order to get at behavioral details and the cognition behind them, we need more than fieldwork. No one would try to measure a child's intelligence by watching him run around in the schoolyard with his friends. Mere observation doesn't offer much of a peek into the child's mind. Instead, we bring the child into a room and present him with a coloring task or a computer game, let her stack wooden blocks, ask questions, and so on. This is how we measure human cognition, and it is also the best way to determine how smart apes are. Fieldwork offers hints and suggestions but rarely allows firm conclusions. One may encounter wild chimpanzees who crack nuts with stones, for example, but it is impossible to know how they discovered this technique or how they learn it from one another. For this, we need carefully controlled experiments on naïve chimpanzees who receive nuts and stones for the first time.

Captive apes under enlightened conditions (such as a sizable group in a spacious outdoor area) have the added advantage of providing a close-up look at naturalistic behavior that one can't get in the field. Here apes can be watched and videotaped much more fully than is possible in the forest, where primates often disappear into the undergrowth or canopy as soon as things get interesting. Fieldworkers are often left to reconstruct events based on fragmented observations. To do so is an art, and they are very good at it, but it falls short of the behavioral detail routinely collected in captivity. If one studies facial expressions, for example, zoomed-in high-definition videos that can be slowed down are essential, which require well-lit conditions rarely encountered in the field.

No wonder the study of social behavior and cognition has fostered integration between captive and fieldwork. The two represent different pieces of the same puzzle. Ideally, we use evidence from both sources to support cognitive theories. Observations in the field have often inspired experiments in the lab. Conversely, observations in captivity—such as the discovery that chimpanzees reconcile after fights—have stimulated observations in the field on the same phenomenon. If, on the other hand, experimental outcomes clash with what is known about a species's behavior in the wild, it may be time to try a new approach.[11]

With regard to the question of animal culture, in particular, captive and fieldwork are now often combined. Naturalists document geographic variation in the behavior of a given species, suggesting a local origin and transmission. But they often cannot rule out alternative accounts (such as genetic variation between populations), which is why we need experiments to determine if habits can spread by one individual watching another. Is the species capable of imitation? If so, this greatly strengthens the case for cultural learning in the field. Nowadays we move back and forth all the time between both sources of evidence.

But all these interesting developments happened long after my observations at Burgers' Zoo. Following Kummer's example, my goal at the time was to spell out what social mechanisms may underlie observed behavior. Apart from triadic awareness, I spoke of divide-and-rule strategies, policing by dominant males, reciprocal deal making, deception, reconciliation after fights, consolation of distressed parties, and so on. I developed such a long list of proposals that I devoted the rest of my career to fleshing them out, at first through detailed observations, but later also experimentally. Proposals take so much less time to make than their verification! The latter can be very instructive, though. One can set up experiments, for example, in which one individual can do another one favors, as we did with our capuchin monkeys, but then add a condition in which the partner can do favors in return. This allows favors to travel in both directions between two parties. We found that monkeys become noticeably more generous if favors can be done mutually than if only one of them has the opportunity.[12] I love this kind of manipulation, since it allows far more solid conclusions about reciprocity than any observational account. Observations never quite clinch the deal the way experiments can.[13]

Even though *Chimpanzee Politics* opened a new agenda for research while introducing Machiavelli's thinking to primatology, I was never quite happy with "Machiavellian intelligence" as a popular label for this field.[14] This term implies an end-justifies-the-means manipulation of others, ignoring a vast amount of social knowledge and understanding that has nothing to do with one-upmanship. When a female chimpanzee resolves a fight between two juveniles over a leafy branch by breaking it into two and handing each youngster a piece, or when an adult male chimpanzee helps an injured, limping mother by picking up her offspring to carry it for her, we are dealing with impressive social skills that don't fit the "Machiavellian" label. This cynical identifier made sense a few decades ago, when all animal (including human) life was customarily depicted as competitive, nasty, and selfish, but over time my own interests have drifted into the opposite direction. I have devoted most of my research to the exploration of empathy and cooperation. The exploitation of others, by using them as "social tools," remains a great topic and is an undeniable aspect of primate sociality, but it is too narrow a focus for the field of social cognition as a whole. Caring relationships, the maintenance of bonds, and attempts to keep the peace are equally worthy of attention.

The intelligence required to effectively deal with social networks may explain why the primate order underwent its remarkable brain expansion. Primates have exceptionally large brains. Dubbed the *Social Brain Hypothesis* by British zoologist Robin Dunbar, the connection with sociality is supported by a relation between a primate's brain size and its typical group size. Primates that live in larger groups generally have larger brains. I always find it hard, though, to separate social and technical intelligence, since many big-brained species are strong in both domains. Even species that hardly handle any tools in the wild, such as rooks and bonobos, may be quite good at it in captivity. It remains true, though, that social challenges have been neglected for too long in discussions of cognitive evolution, which tend to focus on interactions with the environment. Given how all-important social problem solving is in the lives of our subjects, primatologists have been right to amend this view.[15]

Triadic Awareness

Siamangs—large black members of the gibbon family—swing high up in the tallest trees of the Asian jungle. Every morning, the male and female burst into spectacular duets. Their song begins with a few loud whoops, which gradually build into ever louder, more elaborate sequences. Amplified by balloonlike throat sacs, the sound carries far and wide. I have heard them in Indonesia, where the whole forest echoed with their sound. The siamangs listen to one another during breaks. Whereas most territorial animals need only to know where their boundaries run and how strong and healthy their neighbors are, siamangs face the added complexity that territories are jointly defended by pairs. This means that pair-bonds matter. Troubled pairs will be weak defenders, while bonded pairs will be strong ones. Since the song of a pair reflects their marriage, the more beautiful it is, the more their neighbors realize not to mess with them. A close-harmony duet communicates not only "stay out!" but also "we're one!" If a pair duets poorly, on the other hand, uttering discordant vocalizations that interrupt one another, neighbors hear an opportunity to move in and exploit the pair's troubled relationship.[16]

To understand how others relate to one another is a basic social skill that is even more important for group-living animals. They deal with a far greater variety than the siamang. In a baboon or macaque troop, for example, a female's rank in the hierarchy is almost entirely decided by the family from which she hails. Owing to a tight network of friends and kin, no female escapes the rules of the matrilineal order according to which daughters born to high-ranking mothers will themselves become high-ranking, while daughters from families at the bottom will also end up at the bottom. As soon as one female attacks another, third parties move in to defend one or the other so as to reinforce the existing kinship system. The youngest members of the top families know this all too well. Born with a silver spoon in their mouth, they freely provoke fights with everyone around, knowing that even the biggest, meanest female of a lower clan will not be allowed to assert herself against them. The youngster's screams will mobilize her powerful mother and sisters. In fact, it has been shown that screams sound different depending on the kind of opponent a monkey confronts. Thus, it is immediately clear to the entire troop whether a noisy fight fits or violates the established order.[17]

The social knowledge of wild monkeys has been tested by playing the distress calls of a juvenile from a loudspeaker hidden in the bushes at a moment when the juvenile itself is out of sight. Hearing this sound, nearby adults not only look in the direction of the speaker but also peek at the juvenile's mother. They recognize the juvenile's voice and seem to connect it with its mother, perhaps wondering what she is going to do about the trouble her offspring is in.[18] The same sort of social knowledge can be seen at more spontaneous moments, when a juvenile female picks up an infant that is unsteadily walking about, only to carry it back to its mother, which means that she knows which female the infant belongs to.

In white-faced capuchin monkeys, the American anthropologist Susan Perry analyzed how individuals form coalitions during fights. Having followed these hyperactive monkeys for over two decades, Susan knows them all by name and life history. During a visit to her field site in Costa Rica, I saw the characteristic coalition stance firsthand. Known as the *overlord*, two monkeys threaten a third with stares and wide-open mouths, one leaning on top of the other. Their opponent thus faces an intimidating display of two monkeys wrapped into one, with both threatening heads stacked on top of each other. Comparing these coalitions with known social ties, Susan found that capuchins preferentially recruit friends who are dominant over their opponent. This by itself is rather logical, but she also found that instead of seeking the support of

Two white-faced capuchin monkeys adopt an "overlord" position, so that their adversary is confronted by two threatening faces and sets of teeth at once.

their best buddies, they specifically recruit those who are closer to themselves than to their opponent. They seem to realize that there is no point appealing to their opponent's buddies. This tactic, too, requires triadic awareness.[19]

Capuchins solicit support by abruptly jerking their heads back and forth between a potential supporter and their adversary, a behavior known as *headflagging*, which is also used against danger, such as a snake.

In fact, these monkeys threaten everything they don't like, a tendency sometimes used to manipulate attention. Susan once observed the following deceptive sequence:

> *Pursued by a coalition of three higher-ranking males, Guapo suddenly stopped in his tracks and began to produce frantic snake alarm calls while looking at the ground. I was standing by him and could plainly see that there was nothing there but bare ground. He headflagged to Curmudgeon [one of his enemies] for support against the imaginary snake. Guapo's pursuers stopped short and stood up on their hind legs to see if there was a snake. After cautious inspection, they once again began threatening Guapo. Switching tactics he glanced up at a passing magpie jay (a nonmenacing bird) and did three bird alarms in rapid succession—calls that are usually reserved for large raptors and owls. Guapo's opponents looked up, saw that it was not a dangerous bird, and again resumed threatening Guapo. He reverted to the snake alarm call tactic once again vehemently bouncing at the bare patch of ground, threatening the "snake" vocally. Although Curmudgeon continued to glare at Guapo for a bit longer, the rest of the group stopped threatening him, and he was able to resume foraging for insects, moving slowly and nonchalantly towards Curmudgeon while occasionally casting a furtive glance in his direction.[20]*

While such observations suggest but cannot prove high intelligence, there is an urgent need for information on the cognition of wild primates. Fieldworkers are finding ingenious ways to collect it. In Budongo Forest, in Uganda, for example, Katie Slocombe and Klaus Zuberbühler set out to record the screams of chimpanzees under threat or attack. These loud vocalizations serve to recruit aid, which prompted the scientists to see if the acoustics of screams depend on the audience. Given the dispersed lives of wild chimpanzees, only individuals who are within earshot—the audience—are likely to provide aid to a screaming victim. In addition to finding that the intensity of the calls reflected the intensity of the attack, the scientists noted a subtle deception encoded in them. Chimpanzee victims apparently exaggerate their screams (making the attack sound more severe than it truly is), provided their audience includes individuals that outrank their attacker. In other words, whenever the big bosses are around, chimp victims scream bloody murder. Their vocal distortion of the truth suggests precise knowledge of their opponent's status relative to everyone else.[21]

More evidence that primates know one another's relationships comes from the way they classify others based on family membership. Some studies have explored their tendency to *redirect* aggression. Recipients of aggression often look for a scapegoat, not unlike the way people who get reprimanded at work may come home to maltreat their spouse and children. Given their strict hierarchies, macaques are a prime example. As soon as one of these monkeys gets threatened or chased, it will threaten or chase somebody else, always an easy target. Redirected hostility thus travels down the pecking order. Remarkably, redirecting monkeys prefer to target the family of the original aggressor. One monkey will be attacked by a high-ranking individual, then look around to spot a younger, less powerful member of her attacker's family to take her tensions out on this poor soul. This way redirection resembles revenge, since it makes the family of the instigator pay.[22]

The same knowledge of family relations also serves more constructive purposes, such as when after a fight between two monkeys of different families, tensions are resolved by *other* members of the same families. Thus, if play between two juveniles turns into a screaming fight, their mothers may get together to make up for their children. It is an ingenious system, but again it requires every monkey to know to which family every other monkey belongs.[23]

Categorizing others into families may be a case of *stimulus equivalence*, as proposed by the late Ronald Schusterman, an American marine mammal specialist. Ron kept the strangest and most delightful animal laboratory that I have ever set a slippery foot in, since it consisted of not much more than an outdoor swimming pool in sunny Santa Cruz, California. It was the ultimate wet lab. To the side of the pool stood a few wooden panels on which symbols could be mounted for his sea lions. The animals swam in the pool, racing around faster than any human ever could, only to jump out for a few seconds and touch a symbol with their wet noses. Ron's star performer was his favorite pinniped, named Rio. If Rio made the right choice, a fish would be thrown at her, and she'd dive right back into the pool. She did all this in one fluid movement, catching the fish while sliding back into the water, reflecting perfect coordination between experimenter and subject. Ron explained that most tests were too simple for Rio, resulting in her getting bored and losing her concentration. Making errors, she'd get mad at Ron for not giving her enough fish and angrily toss all her plastic toys out of the pool.

Rio had learned to associate arbitrary symbols. She'd first learn that symbol A belongs with B, then that B belongs with C, and so on. After rewarding her for making the right connections, Ron would surprise her with a brand-new combination, such as A and C. If A and B are equivalent as well as B and C, then A and C must be equivalent, too. Would Rio extrapolate from the previous associations, and group A, B, and C together? She did, applying this logic to combinations that she had never encountered before. Ron saw this as the prototype of how animals may mentally group individuals together, such as families or cliques.[24] We do the same: if you learned to connect me first with one of my brothers, then also with another one (I have five!), you should also group those two brothers together in the same family even if you have never encountered them together. Equivalence learning makes for quick and efficient categorization.

Ron went further, speculating about other unseen connections. For example, chimpanzee males have been known to angrily attack and destroy the empty night nests that rival males have left behind in trees at the border of their territory. Unable to attack the enemy itself, the next best target is apparently a nest that they have built. It reminds me of a time in the Netherlands, when owners of black Suzuki Swifts had a tough time. They suffered frequent nasty remarks from people and worse, such as intentional damage to their cars. This situation arose after someone with murderous intentions had driven a black Suzuki Swift into a festive crowd on Queen's Day, killing eight people. The car itself was obviously not at fault, but humans are quick to connect the dots. A hated action turned a specific car brand into a hated object. It all boiled down to stimulus equivalence.

Knowing as we do the spontaneous use of triadic awareness, the next question is how it is acquired. To find out, we need experiments. Is it enough for animals to just watch others? In one study, the French psychologist Dalila Bovet rewarded rhesus monkeys at Georgia State University for identifying the dominant monkey in a video. The observing monkeys didn't know the individuals they were watching and had to judge their relationship purely on the basis of behavior. For example, one monkey in the video would chase another, after which the observer would be trained to select the dominant one (the one who had done the chasing) on a freeze frame of the scene. After learning to do this, the observing monkeys generalized to behaviors that didn't look like chasing but also indicated dominance. Subordinate rhesus monkeys, for example, communicate their position to the dominant by baring their teeth in a wide grin. Bovet showed videos in which this signal was being exchanged. Even though these scenes were new to the observing monkeys, they correctly picked the dominant party. The conclusion was that they have a concept of rank and are quick to evaluate the status of unknown individuals on the basis of how they interact with others.[25]

Ravens may show a similar understanding, as evidenced by their reactions to vocalizations played over a loudspeaker. Ravens recognize one another's voices and pay close attention to dominant and subordinate calls. But then the playbacks were manipulated to make it sound as if a dominant individual had turned submissive. Hearing evidence of a brewing overthrow, the ravens would stop what they were doing and listen while showing signs of distress. They were most upset by rank reversals among members of their own sex in their own group, but they reacted also to status reversals between ravens in an adjacent aviary. The investigators concluded that ravens have a concept of status that goes beyond their own position. They know how others typically interact and are alarmed by deviations from this pattern.[26]

In a related question, I have always wondered if captive chimpanzees evaluate status differences among the people around them. I once worked at a zoo with a demanding director who would occasionally visit the facilities and order everyone around, pointing out problems, saying this needed to be cleaned, and that needed to be moved, and so on. Showing typical alpha conduct, he kept everyone on their toes, as a good director should. Even though the chimps rarely interacted with him— he never fed them or talked to them—they picked up on this behavior. They treated this man with the utmost respect, greeting him with submissive grunts from a great distance (which they didn't do for anyone else) as if they realized, *Here comes the boss, the one everyone around here is nervous about.*

It's not just in relation to dominance that chimpanzees make such judgments. One of the best illustrations of their triadic awareness occurs in mediated conflict resolution. After a fight between male combatants, a third party may induce them to make peace. Interestingly, it's only female chimps who do so, and only the highest-ranking ones among them. They step in when two male rivals fail to reconcile. The male rivals may be sitting near each other and avoiding eye contact, unable or unwilling to make the first move. If a third male were to approach, even to make peace, he'd be perceived as a party to the conflict. Male chimps form alliances all the time, so their presence is never neutral.

This is where the older females come in. The matriarch of the Arnhem colony, Mama, was the mediator par excellence: no male would ignore her or carelessly start a fight that might incur her wrath. She would approach one of the males and groom him for a while, then slowly walk toward his rival while being followed by the first. She would look around to check on the first and return to tug at his arm if he was reluctant. Then she'd sit down next to the second male, while both males would groom her, one on each side. Finally Mama would slip away from the scene, and the males would pant, splutter, and smack more loudly than before—sounds that signal grooming enthusiasm; but by then they would of course be grooming each other.

In other chimpanzee colonies, too, I have seen old females reduce male tensions. It is a risky affair (the males are obviously in a grumpy mood), which is why younger females, instead of trying to mediate themselves, encourage others to do so. They approach the top female while looking around at the males who are refusing to make up. This way, they try to get something going that they can't accomplish safely by themselves. Such behavior demonstrates how much chimpanzees know about the social relationships of others, such as what has happened between the rival males, what has to be done to restore harmony, and who will be the best one to undertake this mission. It is the sort of knowledge that we take for granted in our own species, but without it animal social life could never have reached its known complexity.

Proof in the Pudding

While cleaning out the old library at the Yerkes Primate Center, we unearthed forgotten treasures. One was the old wooden desk of Robert Yerkes, which is now my personal desk. The other was a film that probably had not been looked at for half a century. It took us a while to find the right projector, but it was worth the trouble. Lacking sound, the film had written titles inserted in between poor-quality black-and-white scenes. It featured two young chimpanzees working together on a task. In true slapstick style, befitting the movie's flickering format, one of the chimps would slap the other on her back every time her dedication flagged. I have shown a digitized version to many audiences, causing much laughter in recognition of the humanlike encouragements. People are quick to grasp the movie's essence: apes have a solid understanding of the advantages of cooperation.

The experiment was run in the 1930s by Meredith Crawford, a student of Yerkes.[27] We see two juveniles, Bula and Bimba, pulling at ropes attached to a heavy box outside their cage. Food has been placed on the box, which is too heavy for one of them to pull in alone. The synchronized pulling by Bula and Bimba is remarkable. They do so in four or five bursts, so well coordinated that you'd almost think they were counting—"one, two, three . . . pull!"—but of course they are not. In a second phase, Bula has been fed so much that her motivation has evaporated, and her performance is lackluster. Bimba solicits her every now and then, poking her or pushing her hand toward the rope. Once they have successfully brought the box within reach, Bula barely collects any food, leaving it all to Bimba. Why did Bula work so hard with so little interest in the payoff? The likely answer is reciprocity. These two chimps know each other and probably live together, so that every favor they do for each other will likely be repaid. They are buddies, and buddies help each other out.

This pioneering study contains all the ingredients later expanded upon by more rigorous research. The *cooperative pulling paradigm*, as it is known, has been applied to monkeys, hyenas, parrots, rooks, elephants, and so on. The pulling is less successful if the partners are prevented from seeing each other, so success rests on true coordination. It is not as if the two individuals pull at random and, by luck, happen to pull together.[28] Furthermore, primates prefer partners who cooperate eagerly and are tolerant enough to share the prize.[29] They also understand that a partner's labor requires repayment. Capuchin monkeys, for example, seem to appreciate each other's effort in that they share more food with a partner who has helped them obtain it than with one whose help went unneeded.[30] Given all this evidence, one wonders why the social sci-

ences in recent years have settled on the curious idea that human cooperation represents a "huge anomaly" in the natural realm.[31]

It has become commonplace to assert that only humans truly understand how cooperation works or know how to handle competition
and freeloading. Animal cooperation is presented as mostly based on
kinship, as if mammals were social insects. This idea was quickly dis-
proven when fieldworkers analyzed DNA extracted from the feces of
wild chimpanzees, which allowed them to determine genetic relatedness. They concluded that the vast majority of mutual aid in the forest
occurs between unrelated apes.[32] Captive studies have shown that even
strangers—primates who didn't know each other before they were put
together—can be enticed to share food or exchange favors.[33]

Despite these findings, the human uniqueness meme keeps stubbornly replicating. Are its proponents oblivious to the rampant, varied,
and massive cooperation found in nature? I just attended a conference
on Collective Behavior: From Cells to Societies, which addressed the
extraordinary ways in which single cells, organisms, and entire species
realize goals together.[34] Our best theories about the evolution of cooperation stem from the study of animal behavior. Summarizing these ideas
in his 1975 book *Sociobiology*, E. O. Wilson helped launch the evolutionary approach to human behavior.[35]

Excitement about Wilson's grand synthesis seems to have faded,
though. Perhaps it was too sweeping and inclusive for disciplines that
consider humans in isolation. Chimpanzees in particular are nowadays
often depicted as so aggressive and competitive that they can't be truly
cooperative. If this applies to our closest relatives, so the thinking goes,
we can justifiably ignore the rest of the animal kingdom. One prominent
advocate of this position, the American psychologist Michael Tomasello,
extensively compared children and apes, which has led him to conclude
that our species is the only one capable of shared intentions in relation
to common goals. He once condensed his view in the catchy statement

At Burgers' Zoo, live trees are surrounded by electrified wire, yet the chimps manage to get into them anyway. They break long branches out of dead trees and carry them to a live one, where one of them holds the branch steady while another scales it.

"It is inconceivable that you would ever see two chimpanzees carrying a log together."[36]

This is quite an assertion, given Emil Menzel's photographed and filmed sequences of juvenile apes recruiting one another to collectively prop a heavy pole up against the wall of their enclosure in order to get out.[37] I have regularly seen chimps use long sticks as ladders to get across hot wire surrounding live beech trees; one chimp holds the stick while another scales it to reach fresh leaves without getting shocked. We have also videotaped two adolescent females who regularly tried to reach the window of my office, which overlooks the chimp compound at the Yerkes Field Station. Both females would exchange hand gestures while moving a heavy plastic drum right underneath my window. One ape would jump onto the drum, after which the other would climb on top of her and stand on her shoulders. The two females would then synchronously bob up and down like a giant spring; the one standing on top would reach for my window every time she came close. Well synchronized and clearly of the same mind, these females played this game often in alternating roles. Since they never succeeded, their common goal was largely imaginary.

Literally carrying a log together may not be part of these efforts, but this behavior is trained for all the time in Asian elephants. Until recently, the forest industry in Southeast Asia employed elephants as beasts of burden; now they are rarely used for this purpose anymore, but they still demonstrate their skills for tourists. At the Elephant Conservation Center near Chiang Mai, in Thailand, two tall adolescent bulls will effortlessly pick up a long log with their tusks, each standing on one end, draping their trunks over the log to keep it from rolling off. Then they will walk in perfect unison several meters apart, with the log between them, while the two mahouts on their necks sit chatting and laughing and looking around. They are most certainly not directing every move.

Training is obviously part of this picture, but one cannot train any animal to be so coordinated. One can train dolphins to jump synchronously because they do so in the wild, and one can teach horses to run together at the same pace because wild horses do the same. Trainers build on natural abilities. Obviously, if one elephant were to walk slightly faster than the other while carrying the log, or hold it at the wrong height, the whole enterprise would quickly unravel. The task requires step-by-step harmonization of rhythm and movement by the bulls themselves. They have moved from an "I" identity (I perform this task) to a "we" identity (we do this together), which is the hallmark of collective action. They end their performance by lowering the log together, moving it from their tusks into their trunks and then slowly to the ground. They set the heaviest log down on a pile without a single sound, impeccably coordinated.

When Josh Plotnik tested elephants on the cooperative pulling paradigm, he found a solid understanding for the need to synchronize.[38] Teamwork is even more typical of group hunters, such as humpback whales, which blow hundreds of bubbles around a school of fish; the column of bubbles traps the fish like a net. The whales act together to make the column tighter and tighter, until several of them surface through its center with mouths wide open to swallow the bounty. Orcas go even further, in an action so astonishingly well coordinated that few species, including humans, would be able to match it. When orcas along the Antarctic Peninsula spot a seal on an ice floe, they reposition the floe. It takes lots of hard work, but they push it out into open water. Then four or five whales line up side by side, acting like one giant whale. They rapidly swim in perfect unison toward the floe, creating a huge

The highest level of joint intentionality in the animal kingdom is perhaps achieved by killer whales. After spy-hopping to get a good look at a seal on an ice floe, several of them will line up and swim toward the floe at high speed in perfect unison. Their behavior creates a massive wave that washes the seal off the floe straight into some waiting mouths.

wave that washes off the unlucky seal. We don't know how the killer whales agree on the lineup or how they synchronize their actions, but they must be communicating about it before making their move. It is not entirely clear why they do it, because even though the orcas afterward carry the seal around, they often end up releasing it. One seal was deposited back onto a different ice floe to live another day.[39]

On land, lions, wolves, wild dogs, Harris's hawks (teams of which control the pigeons at London's Trafalgar Square), capuchin monkeys, and so on, exhibit plenty of tight teamwork, too. The Swiss primatologist Christopher Boesch has described how chimpanzees hunt colobus monkeys in Ivory Coast: some males act as drivers, while others take up distant positions high up in a tree as ambushers waiting for the monkey troop to escape in their direction through the canopy. Since these hunts take place in the dense jungle of Taï National Park, and both the chimps and the monkeys are dispersed, it is hard to pinpoint what is going on in three-dimensional space, but it appears to involve role division and the anticipation of prey movement. The prey is captured by one of the ambushers, who potentially could quietly slink away with the meat but does exactly the opposite. During the hunt the chimps are silent, but as soon as a monkey is captured, they erupt in a pandemonium of hooting and screaming that draws everyone in, leading to a large cluster of males, females, and young jostling for position. I once stood under a tree (in a different forest) while this happened, and the deafening noise above me left little doubt about how highly chimps prize their meat. Sharing appears to favor hunters over latecomers—even the alpha male may go empty-handed if he failed to participate. The chimpanzees seem to recognize contributions to success. The communal feast that ensues is the only way to sustain this sort of cooperation, because why would anyone invest in a joint enterprise if not for the prospect of a joint payoff?[40]

These observations obviously contradict the view that chimpanzees, and other animals, lack joint action based on shared intentions. One can imagine the head butting between two scientists with such diametrically opposite views as Boesch and Tomasello, who have offices in the same building. Was their appointment as codirectors of the Max Planck Institute in Leipzig an experiment on how human collaboration fares in the face of disagreement? Given these divergent perspectives, let me return to the experiments that led Tomasello to his human uniqueness claim. After testing both children and apes on a cooperative pulling task, he concluded that only the children exhibit shared intentionality.

The question of comparability has come up before, however, and fortunately there are photographs of the respective setups.[41] One shows two apes in separate cages, each with a little plastic table in front of him that he can pull closer with a rope. Oddly, the apes do not occupy a shared space, as in Crawford's classical study. Their cages are not even adjacent: there is distance and two layers of mesh between them—a situation that hampers visibility and communication. Each ape focuses on its own end of the rope, seemingly unaware of what the other is up to. The photo of the children, in contrast, shows them sitting on the carpeting of a large room with no barriers between them. They, too, are using a pulling apparatus, but they sit side by side in full view of each other and are free to move around, touch each other, and talk. These different arrangements go a long way toward explaining why the children showed shared purpose, and the apes did not.

Had this comparison concerned two different species—rats and mice, say—we would never have accepted such dissimilar setups. If rats had been tested on a joint task while sitting side by side and mice while being kept apart, no sensible scientist would permit the conclusion that rats are smarter or more cooperative than mice. We'd demand the same procedure. Comparisons between children and apes get exceptional leeway, however, which is why studies keep perpetuating cognitive differences that, in my mind, are impossible to separate from methodological ones.

In view of the ongoing controversy, we decided to move away from pair-wise testing—whether separate or together—and develop a more naturalistic setup. I sometimes refer to it as our proof-in-the-pudding experiment, since we sought to determine once and for all how well chimps handle conflicting interests: what happens to cooperation in the face of competition? The only way to see which tendency prevails is to provide an opportunity for the chimpanzees to express both at the same time.

My student Malini Suchak came up with the right apparatus to test a colony of fifteen chimps at the Yerkes Field Station. Mounted on the fence of their outdoor enclosure was an apparatus that required very precise coordination to be moved closer to obtain rewards: either two or three individuals had to pull at exactly the same time at separate bars. To coordinate with two partners was harder than with only one, but the apes had no trouble either way. They were sitting spaced out but in full view of one another. Since the whole group was present, there were many possible partner configurations. The apes could decide who to work with while also being on the alert for competitors, such as dominant males or females, as well as freeloaders who might steal rewards without doing any work. They could freely exchange information and freely choose partners, but also freely compete. No large-scale experiment of this kind had ever been tried.

If it is true that chimps can't overcome competition, the test should produce total chaos! The colony should descend into a bickering bunch of apes, fighting over rewards and chasing one another away from the test site. Competitiveness should kill all shared objectives. I knew chimps long enough, however, that I didn't worry much about the outcome of this test; I had studied conflict resolution among them for decades. Despite their poor reputation, I had seen too many scenes of chimpanzees trying to keep the peace and reduce tensions to worry that they would all of a sudden abandon such efforts.

Since Malini and the rest of us wished to see if the chimps could figure out the task on their own, she gave them no pretraining at all. All they knew was that there was a new apparatus and that food was associated with it. They proved remarkably quick learners, realizing that they had to work together and mastering both two-way and three-way pulls within days. Sitting next to one of the pull-bars, Rita would look up at her mother, Borie, who was asleep in a nest on top of a tall climbing frame. She'd climb up all the way to poke Borie in her ribs until she would come down with her. Rita would head for the apparatus, all the while looking over her shoulder to make sure Mom was following. Sometimes we had the impression that the chimps had reached an agreement without us knowing how. Two of them would walk side by side out of the night building, which is quite a distance away, and together head straight for the apparatus, as if they knew exactly what they were going to do. Talk about shared intentionality!

The main point of the study was to see if the apes would compete or cooperate. Clearly, cooperation won big time. We saw some aggression but virtually no injuries. Most fights were low level, such as pulling at someone to drag him or her away from the apparatus, chasing someone off, or throwing sand. Individuals also tried to gain access by grooming one of the pullers until this individual allowed them to take their spot. Cooperation at the apparatus went on almost nonstop, resulting in a total of 3,565 joint pulls.[42] Freeloaders were avoided and occasionally punished for their activities, while overly competitive individuals quickly found out how unpopular their behavior made them. The experiment was conducted over many months, affording plenty of time for all the chimps to learn that tolerance paid off in terms of finding partners to work with. In the end, we found proof in the pudding that chimpanzees are highly cooperative. They have no trouble whatsoever regulating and dampening strife for the sake of achieving shared outcomes.

One possible reason that the behavior we observed was more in line with what is known from the natural habitat may be our colony's background: by the time we tested them, our chimps had lived together for almost four decades. This is a long time by any standard, resulting in an unusually well-integrated group. But when we recently tested a newly formed group, in which many individuals had known one another for only a few years, we found the same high level of cooperation and low level of aggression. In other words, chimpanzees are generally good at conflict management for the sake of cooperation.

The current reputation of chimpanzees as violent and belligerent—"demonic" even—is almost entirely based on the way they treat members of neighboring groups in the wild: they occasionally carry out brutal attacks over territory. This fact has tainted their image, even though lethal combat is so rare that it took decades for scientists to agree on its occurrence. The rate of fatalities at any given field site is on average once every seven years.[43] Moreover, it is not as if this behavior sets chimpanzees apart from ourselves. So why is it used as an argument against their cooperative nature, whereas in our own species intergroup warfare is rightly viewed as a collective enterprise? The same holds for chimpanzees—they almost never attack neighbors on their own. It is time for us to see them for what they are: talented team players who have no trouble suppressing conflicts within their group.

A recent experiment at the Lincoln Park Zoo in Chicago confirmed their cooperative skills. Scientists let a group of chimpanzees fish with dipsticks for ketchup that was stored in the holes of an artificial "termite" mound. At the beginning of the experiment, there were enough holes for all members to feed independently, but then the number of holes was reduced by one each day, until there were very few left. Since each hole was monopolizable, it was thought that the chimps would compete and squabble over access to the dwindling resource. But nothing of the kind happened. They adjusted to their new situation by doing the exact opposite: they peacefully gathered around the remaining holes—usually two at a time, sometimes in trios—dipping their sticks into them in alternation, each chimp politely awaiting his or her turn. Instead of a rise in conflict, all the scientists observed was sharing and turn taking.[44]

When two or more intelligent, cooperative species meet around food resources, the outcome may also be cooperation rather than competition. Each species knows how to take advantage of the other. Fishing cooperatives, in which humans and cetaceans (whales and dolphins) work together, are probably thousands of years old, having been reported from Australia and India to the Mediterranean and Brazil. In South America they operate on the mud shores of lagoons. Fishermen announce their arrival by slapping the water, upon which bottlenose dolphins emerge to herd mullet toward them. The fishers wait for a signal from the dolphins, such as a distinctive type of dive, to throw their nets. Dolphins also do such herding among themselves, but here they drive the fish toward the fishermen's nets. The men know their dolphin partners individually, having named them after famous politicians and soccer players.

Even more spectacular are the cooperatives between humans and killer whales. When whaling still occurred around Twofold Bay, in Australia, orcas would approach the whaling station to perform conspicuous breaching and lobtailing that served to announce the arrival of a humpback whale. They would herd the large whale into shallow waters close to a whaling vessel, allowing the whalers to harpoon the harassed leviathan. Once the whale was killed, the orcas would be given one day to consume their preferred delicacy—its tongue and lips—after which the whalers would collect their prize. Here too humans gave names to their preferred orca partners and recognized the tit-for-tat that is the foundation of all cooperation, human as well as animal.[45]

There is only one area in which human cooperation goes well beyond what we know of other species: its degree of organization and scale. We have hierarchical structures to set up projects of a complexity and duration not found elsewhere in nature. Most animal cooperation is self-organized in that individuals fulfill roles according to their capacities. Sometimes animals coordinate as if they have agreed on a task division beforehand. We do not know how shared intentions and goals are communicated, but they do not seem to be orchestrated from above by leaders, as in humans. We develop a plan and put a hierarchy in place to manage its execution, which allows us to lay a railroad track across the country or build a huge cathedral that takes generations to complete.

Relying on age-old evolved tendencies, we have shaped our societies into complex networks of cooperation that can take on projects of an unprecedented magnitude.

Fishy Cooperation

Cooperation experiments often ask cognitive questions. Do the actors realize they need a partner? Do they know the partner's role? Are they prepared to share the spoils? If one individual were to hog all the benefits, this obviously would imperil future cooperation. So we assume that animals watch not only what they get but also what they get compared to what their partner gets. Inequity is something to worry about.

This insight inspired an immensely popular experiment that Sarah Brosnan and I conducted with pairs of brown capuchins. After they performed a task, we rewarded both monkeys with cucumber slices and grapes after determining that they all favored the latter over the former. The monkeys had no trouble with the task if they received identical rewards, even if they both got cucumber. But they were vehemently opposed to unequal outcomes, if one got grapes and the other got cucumber. The cucumber monkey would contentedly munch on her first slice, but after noticing that her companion was getting grapes, she would throw a tantrum. She'd ditch her measly veggies and shake the testing chamber with such agitation that it threatened to break apart.[46]

Refusing perfectly fine food because someone else is better off resembles the way humans react in economic games. Economists call this response "irrational," since getting something is by definition better than getting nothing. No monkey, they say, should ever refuse food that she'd normally eat, and no human should reject a small offer. One dollar is still better than no dollar. Sarah and I are unconvinced that this kind of reaction is irrational, though, since it seeks to equalize outcomes, which is the only way to keep cooperation flowing. Apes may even go further than monkeys in this respect. Sarah found that chimpanzees sometimes

An odd couple of hunters: a coral trout and a giant moray eel prowl together around the reef.

protest inequity that goes the other way. They object not only to getting *less* than the other but also to getting *more*. Grape receivers may reject their own advantage! This obviously brings us close to the human sense of fairness.[47]

Without going into further details, something encouraging happened in these studies. They were soon extended to other species, including outside the primates. It is always a sign of a field's maturity when it expands. Researchers who applied inequity tests to dogs and corvids found reactions similar to those of the monkeys.[48] Apparently, no species can escape the logic of cooperation, whether it involves the selection of good partners or the balance between effort and payoff.

The generality of these principles is best illustrated by the work on fish by Redouan Bshary, a Swiss ethologist and ichthyologist. For years Bshary has been enchanting us with observations of the interplay and mutualism between small cleaner wrasses and their hosts, the large fish from which the cleaners nibble away ectoparasites. Each cleaner fish owns a "station" on a reef with a clientele, which come and spread their pectoral fins and adopt postures that offer the cleaner a chance to do its job. In perfect mutualism, the cleaner removes parasites from the client's body surface, gills, and even the inside of its mouth. Sometimes the cleaner is so busy that clients have to wait in queue. Bshary's research consists of observations on the reef but also experiments in the laboratory. His papers read much like a manual for good business practice. For example, cleaners treat roaming fish better than residents. If a roamer and a resident arrive at the same time, the cleaner will service the roamer first. Residents can be kept waiting since they have nowhere else to go. The whole process is one of supply and demand. Cleaners occasionally cheat by taking little bites of healthy skin out of their client. Clients don't like this and jolt or swim away. The only clients that cleaners never cheat are predators, which possess a radical counterstrategy: to swallow them. The cleaners seem to have an excellent understanding of the costs and benefits of their actions.[49]

In a set of studies in the Red Sea, Bshary observed coordinated hunting between the leopard coral trout—a beautiful reddish-brown grouper that can grow to three feet in length—and the giant moray eel. These two species make a perfect match. The moray eel can enter crevices in the coral reef, whereas the trout hunts in the open waters around it. Prey can escape from the trout by hiding in a crevice and from the eel by entering open water, but it cannot get away from the two of them together. In one of Bshary's videos, we see a coral trout and a moray eel swimming side by side like friends on a stroll. They seek each other's company, with the trout sometimes actively recruiting an eel through a curious head shake close to the eel's head. The latter responds to the invitation by leaving its crevice and joining the trout. Given that the two species don't share the prey with each other but swallow it whole, their behavior seems a form of cooperation in which each achieves a reward without sacrificing anything for the other. They are out for their own gain, which they attain more easily together than alone.[50]

The observed role division comes naturally to two predators with different hunting styles. What is truly spectacular is that the entire pattern—two actors who seemingly know what they are going to do and how it will benefit them—is not one we usually associate with fish.

We have lots of cognitively high-level explanations for our own behavior and find it hard to believe that the same might apply to animals with much smaller brains. But lest one think that the fish are showing a simplified form of cooperation, Bshary's recent work challenges this notion. Coral trout were presented with a fake moray eel (a plastic model capable of performing a few actions, such as coming out of a tube) that was able to help them catch fish. The setup followed the same logic as the pulling tests in which chimpanzees recruit help when needed, but not if they can complete the task alone. The trout acted in every way similar to the apes and were equally adept at deciding on their need for a partner.[51]

One way to look at this outcome is to say that chimpanzee cooperation may be simpler than we thought, but another is to say that fish may have a better understanding of how cooperation works than we have been willing to assume. Whether all this boils down to associative learning by the fish remains to be seen; if it does, then any kind of fish should be able to develop this behavior. That seems doubtful, and I agree with Bshary that a species's cognition is tied to its evolutionary history and ecology. Combined with field observations of cooperative hunting between coral trout and moray eels, the experiment suggests a cognition that suits the hunting techniques of both species. Since the trout takes most of the initiatives and decisions, it may all depend on the specialized intelligence of only one species.

These exciting excursions into nonmammals fit the comparative approach that is the hallmark of evolutionary cognition. There is no single form of cognition, and there is no point in ranking cognitions from simple to complex. A species's cognition is generally as good as what it needs for its survival. Distant species that face similar needs may arrive at similar solutions, as also happened in the domain of Machiavellian power strategies. After my discovery of divide-and-rule tactics in chimpanzees, and Nishida's confirmation of their use in the wild, we now have a report on ravens.[52] It is perhaps no accident that it came from a young Dutchman, Jorg Massen, who spent years with the chimps at

Burgers' Zoo before he set out to follow wild ravens in the Austrian Alps. There he observed many separating interventions in which one bird would interrupt a friendly contact between others, such as mutual preening, either by attacking one of them or by inserting itself between them. The intervener gained no direct benefits (there was no food or mating at stake) but did manage to ruin a bonding session between others. Bonds are important to ravens, because as Massen explains, their status depends on them. High-ranking ravens are generally well bonded, whereas the middle category are loosely bonded, and the lowest birds lack special bonds. Since interventions were mostly carried out by well-bonded birds targeting loosely bonded ones, their main goal may have been to prevent the latter from establishing friendships in order to rise in status.[53] This begins to look a lot like chimpanzee politics, which is exactly what one would expect in a large-brained species with a healthy power drive.

Jumbo Politics

We tend to think of elephants as matriarchal, and this is entirely correct. Elephant herds consist of females with young, occasionally followed around by one or two grown bulls eager to mate. The bulls are only hangers-on. It is hard to apply the term *politics* to these herds, since the females are ranked by age, family, and perhaps personality, all of which traits are stable. There isn't much room for the status competition and the opportunistic making and breaking of alliances that marks political strife. For this, we have to go to the males, also in the elephant.

For the longest time, bull elephants have been viewed as loners who travel up and down the savanna and occasionally get behaviorally transformed by the state of *musth*. Jolted by a twentyfold increase in testosterone, a bull changes into a sort of spinach-eating Popeye, a self-confident jerk ready to fight anyone in his path. Not many animals have such a physiological oddball thrown into their social system. But now we learn from the work by American zoologist Caitlin O'Connell in Estosha National Park, in Namibia, that there is more going on. African elephant bulls are far more sociable than assumed. They may not move in herds like the cows—who stick together to keep predators from bringing down their young—but they know one another individually and have leaders, followers, and semipermanent associations.

In some ways, O'Connell's descriptions remind me of primate politics, but at other times they sound odd due to the strange ways elephants communicate. For example, a leading bull wary of another may drop his penis during a butt-jiggling retreat. What is going on here? He is awkwardly walking backward while his penis—which is pretty obvious in an elephant—serves as a signal. Why not retract it at such moments? They drop it in submission, or as O'Connell calls it, "supplication."

On the dominance side as well, their behavior is highly unusual. Here a description of a *musth* display:

> *He was so agitated that he walked over to the place where Greg had previously defecated and performed a dramatic musth display over the offending pile of feces, dribbling urine and curling his trunk over his head, waving his ears and prancing with his front legs in the air, mouth wide open.*[54]

It used to be thought that the older and larger a bull, the higher-ranking he'd be. If so, this system would be rather inflexible. O'Connell, however, documented status reversals. One leading male gradually lost his ability to rally followers. He would fan his ears and emit a let's-go rumble, but no one would pay any heed the way they had done in earlier years. His coalition was falling apart, whereas it previously had shown impressive cohesion. One sign of an intact "boys' club" is that the dominant bull's vocalizations are echoed by the bulls around him. A subordinate's call starts at the moment the dominant's call ends, followed by yet another subordinate, and yet another, resulting in a cascade of repeated calls among the bulls that signal to the rest of the world that they are tight and united.

Elephant coalitions are subtle, and everything these animals do seems a slow-motion movie to the human eye. Sometimes two bulls will deliberately stand right next to each other with ears out, so as to indicate to an opponent that it is time to leave the waterhole. These coalitions dominate the scene, usually arranged around a clear leader. Other bulls come to pay their respect to him, approaching him with outstretched trunk, quivering in trepidation, dipping the tip into his mouth in an act of trust. After performing this tense ritual, the lower-ranking bulls relax as if a burden has been taken off their shoulders. These scenes are reminiscent of how dominant male chimpanzees expect subordinates to crawl in the dust while uttering submissive grunts, not to mention human status rituals, such as kissing the ring of the don, or Saddam Hussein's insistence that his underlings stick their nose under his armpit. Our species is quite creative when it comes to reinforcement of the hierarchy.

We are familiar enough with these processes to recognize them in other animals. As soon as power is based on alliances rather than individual size or force, the door opens to calculated strategies. Given elephant intelligence in other domains, there is every reason to expect pachyderm society to be as complex as that of other political animals.

7 | TIME WILL TELL

What is time? Leave now for dogs and apes! Man has forever!

—Robert Browning (1896)[1]

Judging the gap between two trees, a monkey relies on its memory of past jumps to calculate the next one. Is there a landing spot on the other side? Is it within jumping distance? Can the branch handle its impact? These life-and-death decisions take a great deal of experience to make and show how past and future intertwine in a species's behavior. The past provides the required practice, whereas the future is where the next move will take place. Long-range future orientation is also common, such as when during a drought the matriarch of an elephant herd remembers a drinking hole miles away that no one else knows about. The herd sets out on a long trek, taking days to reach precious water. While the matriarch operates on the basis of knowledge, the rest of the herd operates on the basis of trust. Whether it is a matter of seconds or days, animal behavior is not only goal- but also future-oriented.

So it is curious to me that animals are often thought to be stuck in the present. The present is ephemeral. One moment it is here, the next it is gone. Whether you are a thrush picking up a worm for your chicks in a distant nest or a dog setting out in the morning to patrol your territory and dribble urine at strategic locations, animals have jobs to do, which imply the future. True, most of the time it is the near future, and it remains unclear how aware they are of it. Yet their behavior would make no sense if they lived entirely in the present.

We ourselves consciously reflect on the past and the future, so it was perhaps unavoidable that whether animals do or don't would become a battleground. Isn't consciousness what sets humans apart? Some claim that we are the only ones to actively recall the past and imagine the future, but others have been busy gathering evidence to the contrary. Since no one can prove conscious reflection without verbal reports, the debate skirts subjective experience as something that—at least for now— we can't put our finger on. There has been genuine progress, though, in the exploration of how animals relate to the time dimension. Of all areas of evolutionary cognition, this one is perhaps the most esoteric and the hardest to get a handle on. The terminology shifts regularly, and debates are fierce. For this reason, I have visited two experts to ask them where we currently stand, which opinions will be presented at the end of this chapter.

In Search of Lost Time

Perhaps the controversy started earlier than we think, because in the 1920s an American psychologist, Edward Tolman, bravely and controversially asserted that animals are capable of more than the mindless linking between stimulus and response. He rejected the idea of them as purely incentive-driven. He dared use the term *cognitive* (he was famous for his studies of cognitive maps in maze-learning rats) and called animals "purposive," guided by goals and expectations, both of which reference the future.

While Tolman—in a bow to the suffocating grip of the era's classical behaviorism—shied away from the stronger term *purposeful*, his student Otto Tinklepaugh designed an experiment in which a macaque watched either a lettuce leaf or a banana being placed under a cup. As soon as the monkey was given access, she ran to the baited cup. If she found the food that she had seen being hidden, everything proceeded smoothly. But if the experimenter had replaced the banana with lettuce, the monkey only stared at the reward. She'd frantically look around, inspecting the location over and over, while angrily shrieking at the sneaky experimenter. Only after a long delay would she settle for the disappointing vegetable. From a behaviorist perspective, her attitude was bizarre since animals are supposed to merely connect behavior with rewards, *any* rewards. The nature of the reward shouldn't matter. Tinklepaugh, however, demonstrated that there is more going on. Guided by a mental representation of what she had seen being hidden, the monkey had developed an expectation, the violation of which deeply upset her.[2]

Instead of merely preferring one behavior over another, or one cup over another, the monkey recalled a specific event. It was as if she were saying "Hey, I swear I saw them put a banana under that cup!" Such precise recall of events is known as *episodic memory*, which was long thought to require language, hence to be uniquely human. Animals were thought to be good at learning the general consequences of behavior without retaining any specifics. This position has become shaky, though. Let me give an example that is a bit more striking since it involves a much longer time frame than the monkey experiment.

We once applied a Menzel-type test to Socko, when he was still an adolescent chimpanzee. Through a small window, Socko watched my assistant hide an apple in a large tractor tire in the outdoor enclosure, while the rest of the colony was kept behind closed doors. Then we released the colony, holding Socko back until last. The first thing he did after coming out the door was to climb onto the tire and peek into it, checking on the apple. He left it alone, though, and nonchalantly walked away from the scene. He waited for more than twenty minutes, until everyone was otherwise occupied, and then went to collect the fruit. This was clever, since he might otherwise have lost his prize.

The truly interesting twist came years later, however, when we repeated this experiment. Socko had been tested only once, and we showed the video to a visiting camera crew. But as is typical, the crew trusted its own filming better and insisted on redoing the whole test. By this time Socko was the alpha male and hence could not be used anymore. Being of high rank, he would have had no reason to conceal what he knew about hidden food. So instead we selected a low-ranking female named Natasha and did everything nearly same. We locked up all the chimps and let Natasha watch through the window while we hid an apple. This time we dug a hole in the ground, put the apple into it, and covered it with sand and leaves. We did this so well that afterward we barely knew where we'd put the fruit.

After the others were released, Natasha finally entered the enclosure. We waited anxiously, following her with several cameras. She showed a pattern similar to Socko's and moreover displayed a far better sense of location than we did. She passed slowly over to the precise hiding spot, then returned ten minutes later to confidently dig up the fruit. While she did so, Socko stared at her with apparent surprise. It is not every day that someone pulls an apple out of the ground! I worried that Socko might punish her for snacking right in front of him, but no, Socko ran straight to the tractor tire! He looked into it from several angles, but obviously it was empty. It was as if he had concluded that we were hiding fruit again—and he recalled the exact location we'd used before. This was most remarkable since I am pretty sure Socko had had only one experience of this kind in his whole life, which had occurred five years earlier.

Was this mere coincidence? It is hard to tell based on a single event, but fortunately a Spanish scientist, Gema Martin-Ordas, has been testing out this sort of memory. Working with a large number of chimpanzees and orangutans, she tested the apes on what they remembered of past events. Previously, the apes had been given a task that required them to find the right tool to fetch either a banana or frozen yogurt. The apes had watched tools being hidden in boxes, after which they needed to pick the right box to get a tool for the task. This being easy for apes, all went well. But three years later, after the apes had gone through scores of other events and tests, they all of a sudden encountered the same person, Martin-Ordas, presenting the same setup in the same rooms of the building. Would the presence of the same investigator and situation cue the apes about the challenge they faced? Would they know right away what tool to use and where to look for it? They did, or at least those with previous experience did. Naïve apes did nothing of the kind, thus confirming the role of memory. And not only that, the apes did not hesitate: they solved the problem in a matter of seconds.[3]

Most animal learning is of a rather vague kind, similar to how I have learned to avoid some Atlanta highways at certain times of the day. Having gotten stuck in traffic often enough, I will look for a better, faster route, without any specific memory of what happened on my previous commutes. This is also how a rat in a maze learns to turn one way and not another, and how a bird learns at what time of day to find bread crumbs at my parents' balcony. This kind of learning is all around us. What we deem a special kind, the one at issue here, is the recall of particulars, the way the French novelist Marcel Proust, in *In Search of Lost Time* dwelled on the taste of a *petite madeleine*. The little tea-soaked biscuit made him relive his childhood visits to Aunt Leonie: "No sooner had the warm liquid mixed with the crumbs touched my palate than a shudder ran through me and I stopped, intent upon the extraordinary thing that was happening to me."[4] The power of autobiographical memories lies in their specificity. Colorful and alive, they can be actively called up and dwelled upon. They are reconstructions—which is why they are sometimes false—yet so powerful that they are accompanied by an extraordinary sense of their correctness. They fill us with emotions and sensations, as happened to Proust. You mention someone's wedding day, or Dad's funeral, and all sorts of memories about the weather, the guests, the food, the happiness, or the sadness will flood the mind.

This kind of memory must be at work when apes react to cues connected to events from years back. The same memory serves foraging wild chimpanzees, which visit about a dozen fruit-bearing trees per day. How do they know where to go? The forest has far too many trees to go about it randomly. Working in Taï National Park, in Ivory Coast, the Dutch primatologist Karline Janmaat found apes to have an excellent recall of previous meals. They mostly checked trees at which they had eaten in previous years. If they ran into copious ripe fruit, they'd gorge on it while grunting contentedly and make sure to return a couple of days later.

Janmaat describes how the chimps would build their daily nests (in which they sleep for only one night) en route to such trees and get up before dawn, something they normally hate to do. The intrepid primatologist followed the traveling party on foot, but whereas the chimps typically ignored her tripping or stepping on a noisy branch, now they all would turn around and stare pointedly at her, making her feel bad. Sounds draw attention, and the chimps were on edge in the dark. This was understandable since one of the females had recently lost her infant to a leopard.

Despite their deep-seated fear, the apes would set out on a long trek to a specific fig tree where they had recently eaten. Their goal was to beat the early fig rush. These soft, sweet fruits are favored by many forest animals, from squirrels to flocks of hornbills, so that an early arrival would be the only way to take advantage of the abundance. Remarkably, the chimps would get up earlier for trees far from their nests than for those nearby, arriving at about the same time at both. This suggests calculation of travel time based on expected distances. All this makes Janmaat believe that the Taï chimpanzees actively recall previous experiences in order to plan for a plentiful breakfast.[5]

The Estonian-Canadian psychologist Endel Tulving defined *episodic memory* as the recall of what happened at which place and at what time. This has prompted research into memory of the three *W*'s of events: their what, when, and where.[6] While the above ape examples seem to fit the bill, we need more tightly controlled experiments. The first challenge to Tulving's claim that episodic memory is limited to humans came from precisely such an experiment, not on apes, but on birds. Together with Anthony Dickinson, Nicky Clayton took advantage of the hoarding tendency of her western scrub jays to see what they remembered about cached foods. The birds were given different items to hide, some perishable (waxworms), others durable (peanuts). Four hours later the jays looked for the worms—their favorite food—before they looked for nuts, but five days later their response was reversed. They didn't even bother to find the worms, which by that time would have spoiled and become distasteful. They did remember the peanut locations after this long interval, though. Odor could be ruled out as a factor, because by the time they were tested, the scientists recorded search patterns in the absence of food. This study was quite ingenious and included a few additional controls, leading the authors to conclude that jays recall what items they have put where and at what point in time. They remembered the three *W*'s of their actions.[7]

The case for episodic memory in animals was further strengthened when the American psychologists Stephanie Babb and Jonathon Crystal let rats run around in an eight-armed radial maze. The rodents learned that once they had visited an arm and eaten the food in it, it would be permanently gone, so there would be no point returning to it. There was one exception, though. They occasionally found chocolate-flavored pellets, which would be replenished after long time intervals. The rats formed an expectation about this delicious food based on where and when they had encountered it. They did return to those specific arms, but only after long intervals. In other words, the rodents kept track of the when, what, and where of chocolate surprises.[8]

Tulving and a few other scholars were hardly satisfied with these results, however. They fail to tell us—the way Proust did so eloquently— how aware the birds, rats or apes are of their own memories. What kind of consciousness, if any, is involved? Do they view their past as a piece of personal history? Since such questions are unanswerable, some have weakened the terminology by endowing animals only with "episodic-like" memory. I don't agree with this retreat, however, since it gives weight to an ill-defined aspect of human memory known only through introspection and language. While language is helpful to communicate memories, it is hardly what produces them. My preference would be to turn the burden of proof around, especially when it comes to species close to us. If other primates recall events with equal precision as humans do, the most economic assumption is that they do so in the same way. Those who insist that human memory rests on unique levels of awareness have their work cut out for them to substantiate such a claim.

It may, literally, be all in our heads.

The Cat's Umbrella

The debate about how animals experience the time dimension heated up even further in relation to the future. Who'd ever heard of them contemplating events that lay ahead? Tulving drew on what he knew about Cashew, his cat. Cashew seems capable of predicting rain, he said, and is good at finding places to take cover, yet "never thinks ahead and packs an umbrella."[9] Generalizing this astute observation to the entire animal kingdom, the eminent scientist explained that while animals adapt to their present environment, they sadly fail to imagine the future.

Another human uniqueness proponent noted that "there is no obvious evidence that animals have ever agreed on a five-year plan."[10] True, but how many humans have? I associate five-year plans with central government and prefer examples drawn from the way both humans and animals go about their daily business. For example, I may plan to buy groceries on my way home, or decide to surprise my students with a quiz next week. This is the nature of our planning. It is not unlike the story with which I opened this book regarding Franje, the chimpanzee who gathered all the straw from her night cage to build a warm nest outdoors. That she took this precaution while still indoors, before actually feeling the cold outside, is significant because it fits Tulving's so-called spoon test. In an Estonian children's story, a girl dreams of a friend's chocolate pudding party where she can only watch other children eat, because everyone has brought their own spoon, and she has not. To prevent this from happening again, she goes to bed that night clutching a spoon. Tulving proposed two criteria to recognize future planning. First, the behavior should not follow directly from present needs and desires. Second, it should prepare the individual for a future situation in a different context than the current one. The girl needed a spoon not in bed, but at the chocolate pudding party she expected in her dream.[11]

When Tulving came up with the spoon test, he wondered if it was perhaps unfair. Wasn't it too demanding for animals? He proposed this test in 2005, well before most experiments on future planning were conducted, apparently unaware that apes pass the spoon test every day in their spontaneous behavior. Franje did so when she gathered straw in a different location and under different circumstances than where it was needed. At the Yerkes Primate Center, we also have a male chimp, Steward, who never enters our testing room without first looking around outdoors for a stick or branch that he uses to point at the various items in our experiments. Even though we have tried to discourage this behavior, by removing the stick from his hands so that he'll point with a finger like everyone else, Steward is stubborn. He prefers to point with a stick and will go out of his way to bring one with him, thus anticipating our test and his self-invented need for a tool.

But perhaps the nicest illustration, out of dozens I could offer, is a bonobo named Lisala, who lives at Lola ya Bonobo, a jungle sanctuary

Lisala, a bonobo, carries a heavy rock on a long trek toward a place where she knows there are nuts. After collecting the nuts, she continues her trek to the only large slab of rock in the area, where she employs her rock as a hammer to crack the nuts. Picking up a tool so long in advance suggests planning.

near Kinshasa where we conducted studies of empathy. The observation in question was unrelated to this topic, however, and was made by my coworker Zanna Clay when she unexpectedly saw Lisala pick up an enormous fifteen-pound rock and lift it onto her back. Lisala carried this heavy load on her shoulders while her baby clung to her lower back. It was rather ridiculous, of course, since it impeded her travel and required extra energy. Zanna turned on her video camera and followed the bonobo to see what the rock might be for. Like any true ape expert, she immediately assumed that Lisala had a goal in mind, because, as Köhler had noted, ape behavior is "unwaveringly purposeful." The same holds for human behavior. If we see a man walking in the street with a ladder, we automatically assume that he wouldn't be carrying such a heavy tool for no reason.

Zanna filmed Lisala's trek of about half a kilometer. It was interrupted only once when she put down the rock and picked up some items that were hard to identify. Then she put the rock back onto her back and continued her travels. She walked all told almost ten minutes before she reached her destination, which was a large slab of hard rock. She cleared it of debris with a few swipes of her hand, then put down her rock, her infant, and the collected items, which turned out to be a handful of palm nuts. She set out to crack these extremely tough nuts, placing them on the large anvil while banging them with her fifteen-pound rock as a hammer. She spent about fifteen minutes on this activity, then left her tool behind. It is hard to imagine that Lisala had gone through all this trouble without a plan, which she must have had well before she picked up the nuts. She probably knew where to find those, hence planned her route via this location, to end up at a point that she knew had a hard enough surface for successful cracking. In a nutshell, Lisala fulfilled all of Tulving's criteria. She picked up a tool to be used at a distant location for the processing of food that she could only have imagined.

Another remarkable instance of future-oriented behavior was documented at a zoo by the Swedish biologist Mathias Osvath, this time involving a male chimpanzee, Santino. Every morning before visitors arrived, Santino would leisurely collect rocks from the moat surrounding his enclosure, stacking them up in neat little piles hidden from view. This way he'd have an arsenal of weapons when the zoo opened its gates. Like so many male chimps, Santino would several times a day rush around with all his hair on end to impress the colony and the public. Throwing stuff around was part of the show, including projectiles aimed at the watching masses. Whereas most chimps find themselves empty-handed at the critical moment, Santino prepared his rock piles for these occasions. He did so at a quiet time of the day, when he was not yet in the adrenaline-filled mood to produce his usual spectacle.[12]

Such cases deserve attention since they show that apes do not have to be prompted by experimental conditions concocted by us humans to plan for the future. They do so of their own accord. Their accomplishments are quite different from the way many other animals orient to upcoming events. We all know that squirrels collect nuts in the fall and hide them for retrieval in winter and spring. Their hoarding is triggered by the shortening of day length and the presence of nuts, regardless of whether the animals know what winter is. Young squirrels naïve about the seasons do exactly the same. Whereas this activity does serve future needs and requires quite a bit of cognition regarding what nuts to store and how to find them again, the seasonal preparations of squirrels are unlikely to reflect actual planning.[13] It is an evolved tendency found in all members of their species and limited to only one context.

The planning of apes, in contrast, adjusts to the circumstances and is flexibly expressed in myriad ways. That it is based on learning and understanding is hard to prove from observation alone, however. It requires subjecting apes to conditions that they have never met before. What happens, for example, if we create a situation in which clutching a spoon, so to speak, is advantageous later on?

The first such study was conducted in Germany by Nicholas Mulcahy and Josep Call, who let orangutans and bonobos select a tool that they couldn't use right away even though the rewards were visible. The apes were moved away to a waiting room to see if they would hold on to their tool for later use even if the right occasion would arise only fourteen hours later. The apes did so, yet it could be (and has been) argued that they might have developed positive associations with certain tools, hence valued them regardless of what they knew about the future.[14]

This issue was addressed by a similar experiment in which apes selected tools, but this time the rewards were kept out of sight. The apes preferred a tool they could use in the future over a grape placed right next to it. They suppressed their desire for an immediate benefit to gamble on a future one. Once they had the right tool in hand, however, and got a second presentation of the same set of tools, they did pick the grape. Clearly, they didn't value the tool over anything else, because if they did, their second choice should have replicated the first. The apes must have realized that once they had the right tool in hand, there was no point having a second one of the same kind, and that the grape was a better choice.[15]

These clever experiments were foreshadowed by Tulving's proposal as well as by Köhler, who was the first to speculate about future planning in animals. There is now even a test in which, instead of presenting apes with actual tools, they are given an opportunity to fabricate them in advance. Apes learned to break a board of soft wood into smaller pieces to produce sticks with which they could reach grapes. Anticipating the need for sticks, they worked hard on having them ready in time.[16] Their preparations resembled the behavior of wild apes, which travel long distances with raw materials that they turn on the spot into tools by modifying, sharpening, or fraying them. They sometimes bring more than one type of tool to a task in the forest. Chimps carry toolkits of up to five different sticks and twigs to hunt for underground ants or raid bee nests for honey. It is hard to imagine an ape searching for and traveling with multiple instruments without a plan. Just so, Lisala picked up a heavy rock that by itself was useless and that could serve its purpose only in combination with nuts that she had yet to collect as well as a hard surface located far away. Attempts to explain this kind of behavior without foresight invariably sound cumbersome and far-fetched.

The question now is whether similar evidence can be produced without reliance on tools such as spoons, umbrellas, or sticks. What if we consider a wider spectrum of behavior? How this might be done was again demonstrated by Clayton's scrub jays. These birds routinely cache food, and although some scientists complain that this behavior offers a rather narrow window on cognition, it is a window nonetheless and one that differs radically from the one used for primates. It exploits an activity that corvids are particularly good at, just as tool studies exploit specialized primate skills. The outcome has been most remarkable.

Caroline Raby offered jays an opportunity to store food in two compartments of their cage that would be closed off during the night. The next morning they would get a chance to visit only one of the two compartments. One compartment had become associated with hunger, since the birds had spent mornings there without breakfast. The second compartment, on the other hand, was known as the "breakfast room" because it was stocked with food every morning. Given a chance in the evening to cache pine nuts, the birds put three times as many nuts in the first room as in the second, thus anticipating the hunger they might suffer there. In another experiment, the birds had learned to associate both compartments with different kinds of food. Once they knew what kind to expect, they tended to store a *different* food in each compartment in the evening. This guaranteed a more varied breakfast if they ended up in one of those compartments next morning. All in all, when scrub jays stash away food, they do not seem guided by their present needs and desires but rather by the ones they anticipate in the future.[17]

In thinking of primate examples without tools, the ones that come to mind are social situations in which it helps to be diplomatic. For example, chimpanzees sometimes arrange a secret rendezvous with the opposite sex. Bonobos don't need to do so, since others rarely interrupt their sexual escapades, but chimpanzees are far less tolerant. High-ranking males don't allow rivals near females with an attractive genital swelling. Nevertheless, the alpha male cannot always be awake and alert, hence occasions do arise for young males to invite a female to get away to a quiet spot. Typically, the young male spreads his legs to show his erection—a sexual invitation—making sure that his back is turned to the other males or that, with his underarm leaning on his knee, one of his hands loosely dangles right next to his penis so that only the wooed female can see it. After this display, the male nonchalantly wanders off in a given direction and sits down out of view of dominant males. Now it is up to the female, who may or may not follow. So as to give nothing away, she usually takes off in a different direction, only to arrive, via a detour, at the same spot as the young male. What a coincidence! The two of them then engage in a quick copulation, making sure to stay silent. It all gives the impression of a well-planned arrangement.

Even more striking are the tactics of adult males challenging each other for status. Given that confrontations are almost never decided between two rivals on their own but involve support for one or the other by third parties, it is to their advantage to influence public opinion beforehand. The males commonly groom high-ranking females or one of their male buddies before launching into a display, with all their hair on end, to provoke a rival. The grooming gives the impression of them currying favors in advance, knowing full well what the next step will be. In fact, there has been a systematic study on this issue. At Chester Zoo in the United Kingdom, Nicola Koyama recorded for over two thousand hours who groomed whom in a large chimpanzee colony. She also noted what kinds of conflicts arose among the males, and who allied with whom. When she compared records on both behaviors—grooming and alliances—from one day to the next, she discovered that males received more support from the individuals they had groomed the day before. This is the sort of tit-for-tat that we are used to in chimpanzees. But since this connection held only for the aggressors, and not for their victims, the explanation was not simply that grooming promotes support. Koyama viewed the connection as part of an active strategy. Males know beforehand which confrontations they are going to incite, and they pave the way for them by grooming their friends a day in advance. This way they make sure to have their backing.[18] It reminds me of the politics at university departments, where colleagues come to my office in the days leading up to an important faculty meeting to influence my vote.

Observations are suggestive yet rarely conclusive. They do, however, give an idea under what circumstances future planning might be useful. If naturalistic observations and experiments point in the same direction, we must be on the right track. For example, a recent study suggested that wild orangutans communicate future travel routes. Orangutans are such loners that their encounters in the canopy have been described as ships passing in the night. They often travel on their own, accompanied only by their dependent offspring, and remain visually isolated for long stretches of time. Auditory information about one another's where-abouts is often all they have.

Carel van Schaik—a Dutch primatologist who once was a fellow student of mine and whose field site on Sumatra I visited—followed wild males right before they went to bed in their self-made nests high up in the trees. He recorded over a thousand whooping calls made by these males before nightfall. These loud calls may last for up to four minutes, and all orangs around pay close attention, because the dominant male (the only fully grown male with well-developed cheek pads, or flanges) is a figure to be reckoned with. There is usually only one such male in a given area of the forest.

Carel found that the direction in which adult males call before going to sleep predicts their travel path the next day. The calls contain this information even if the direction changes from day to day. Females adjust their own routes to the male's, such that sexually receptive females may approach him, and other females know where to find him in case they are being harassed by adolescent males. (Female orangutans generally prefer the dominant male.) Although Carel recognizes the limitations of a field study, his data imply that orangutans know where they will be going and vocally announce their plan at least twelve hours before its execution.[19]

Neuroscience may one day resolve how planning takes place. The first hints are coming from the hippocampus, which has long been known to be vital both for memory and for future orientation. The devastating effects of Alzheimer's typically begin with degeneration of this part of the brain. As with all major brain areas, however, the human hippocampus is far from unique. Rats have a similar structure, which has been intensely studied. After a maze task, these rodents keep replaying their experiences in this brain region, either during sleep or sitting still while awake. Using brain waves to detect what kind of maze paths the rats are rehearsing in their heads, scientists found that more is going on than a consolidation of past experiences. The hippocampus seems also engaged in the exploration of maze paths that the rats have not (yet) taken. Since humans, too, show hippocampal activity while imagining the future, it has been suggested that rats and humans relate to the past, present, and future in homologous ways.[20] This realization, as well as the accumulated primate and bird evidence for future orientation, has swayed the opinion of several skeptics, who used to think that only humans show mental time travel. We are moving ever closer to Darwin's continuity stance, according to which the human-animal difference is one of degree, not kind.[21]

Animal Willpower

A French politician accused of sexual assault was said to have acted like a "randy chimpanzee."[22] How insulting—to the ape! As soon as humans let their impulses run free, we rush to compare them with animals. But as the above descriptions show, rather than give in to sexual desires, chimps have sufficient emotional control to either refrain from them or to arrange privacy first. It all boils down to the social hierarchy, which is one giant behavioral regulator. If everyone were to act the way they wanted, any hierarchy would fall apart. It is built on restraint. Since social ladders are present in species from fish and frogs to baboons and chickens, self-control is an age-old feature of animal societies.

A famous anecdote comes from the early days in Gombe Stream, when chimpanzees still received bananas from humans. The Dutch primatologist Frans Plooij observed an adult male approach the feeding box, which humans could unlock from a distance. Each individual chimpanzee had been put on a strict quota. The unlocking mechanism made a distinctive click, which announced the availability of fruits. But alas, at the very moment that this male heard the click and got lucky, a dominant male appeared on the scene. What to do now? The first male acted as if nothing were the matter. Rather than open the box—and lose his bananas—he sat down at a distance. No dummy either, the dominant male strolled away from the scene. But as soon as he was out of sight, he peeked around a tree trunk to see what the first male was up to. He thus noticed that the other opened the box and quickly relieved him of his prize.

One reconstruction of this sequence is that the dominant male got suspicious since he felt that the other was acting odd. Hence his decision to keep an eye on him. Some have even suggested multiple layers of intentionality: first, that the dominant male suspected that the first male was trying to give the impression that the lid was still locked; second, that the dominant let the other think that he hadn't noticed.[23] If true, this would be a deceptive mind game more complex than most experts are willing to give apes credit for. For me, however, the interesting part is the patience and restraint both males showed. They suppressed the impulse to open the box in each other's presence, even though it contained a highly desirable food that was rarely available.

It is easy to see inhibitions at work in our pets, such as a cat who spots a chipmunk. Instead of going after the little rodent right away, she makes a wide detour, with her body sleekly pressed against the ground, to arrive at a hiding spot from which she can pounce on her unsuspecting prey. Or take the big dog who lets puppies jump all over him, bite his tail, and disturb his sleep without a single growl of protest. While restraint is apparent to anyone in daily contact with animals, Western thought hardly recognizes the ability. Traditionally, animals are depicted as slaves of their emotions. It all goes back to the dichotomy of animals as "wild" and humans as "civilized." Being wild implies being undisciplined, crazy even, without holding back. Being civilized, in contrast, refers to exercising the well-mannered restraint that humans are capable of under favorable circumstances. This dichotomy lurks behind almost every debate about what makes us human, so much so that whenever humans behave badly, we call them "animals."

Desmond Morris once told me an amusing story to drive this point home. At the time Desmond was working at the London Zoo, which still held tea parties in the ape house with the public looking on. Gathered on chairs around a table, the apes had been trained to use bowls, spoons, cups, and a teapot. Naturally, this equipment posed no problem for these tool-using animals. Unfortunately, over time the apes became too polished and their performance too perfect for the English public, for whom high tea constitutes the peak of civilization. When the public tea parties began to threaten the human ego, something had to be done. The apes were retrained to spill the tea, throw food around, drink from the teapot's spout, and pop the cups into the bowl as soon as the keeper turned his back. The public loved it! The apes were wild and naughty, as they were supposed to be.[24]

In line with this misconception, the American philosopher Philip Kitcher labeled chimpanzees "wantons," creatures vulnerable to whichever impulse hits them. The maliciousness and lasciviousness usually associated with this term was not part of his definition, which focused on a disregard of behavioral consequences. Kitcher went on to speculate that somewhere during our evolution we overcame this wantonness, which is what made us human. This process started with "awareness that certain forms of projected behavior might have troublesome results."[25] This awareness is key indeed but is obviously present in lots of animals, otherwise they'd run into all sorts of problems. Why do migrating wildebeest hesitate so long before jumping into the river they seek to cross? Why do juvenile monkeys wait until their playmate's mother has moved out of sight before starting a fight? Why does your cat jump onto the kitchen counter only when you aren't looking? Awareness of troublesome results is all around us.

Behavioral inhibitions have rich ramifications, which extend to the origins of human morality and free will. Without impulse control, what would be the point of distinguishing right from wrong? The philosopher Harry Frankfurt defines a "person" as someone who does not just follow his desires but is aware of them and capable of wishing them to be different. As soon as an individual considers the "desirability of his desires," he becomes a person with freedom of will.[26] But while Frankfurt believes that animals and young children don't monitor or judge their own desires, science is increasingly testing out this very capacity. Experiments on *delayed gratification* present apes and children with a temptation that they need to actively resist for the sake of future gain. Emotional control and future orientation are key, with free will not far behind.

Most of us have seen the hilarious videos of children sitting alone behind a table desperately trying *not* to eat a marshmallow—secretly licking it, taking tiny bites from it, or looking the other way so as to avoid temptation. It is one of the most explicit tests of impulse control. The children have been promised a second marshmallow if they leave the first one alone while the experimenter is away. All they have to do is postpone gratification. But in order to do so, they have to go against the general rule that an immediate reward is more appealing than a delayed one. This is why we find it hard to save money for a rainy day, and why smokers find a cigarette more appealing than the prospect of lasting health. The marshmallow test measures how much weight children assign to the future. Children vary greatly on how well they do on the test, and their success predicts how they will fare later in life. Impulse control and future orientation are a major part of success in society.

Many animals have trouble with a similar task and don't hesitate to eat food right away, probably because in their natural habitat they might otherwise lose it. For other species, delay of gratification is very modest, such as in a recent experiment with capuchin monkeys. The monkeys saw a large rotating plate, like a lazy Susan, featuring one piece of carrot and one piece of banana. Capuchins favor the second food. They first saw one and a little later the second item move by, while sitting behind a window through which they were allowed to reach only once. The majority of monkeys ignored the carrot, letting it pass right in front of them, to hold out for the better reward. Even though the delay between the two was a mere fifteen seconds, they showed enough restraint to consume considerably more banana than carrot.[27] Some species, however, show dramatic control that is more in line with that of our own. For example, a chimpanzee patiently stares at a container into which falls a candy every thirty seconds. He knows he can disconnect the container at any moment to swallow its contents but also that this will stop the candy flow. The longer he waits, the more candies he will gather. Apes do about as well as children on this task, delaying gratification for up to eighteen minutes.[28]

Similar tests have been conducted with large-brained birds. We may not consider birds to need self-restraint, but think again. Many birds pick up food for their young that they could easily swallow themselves. In some species, males feed their mates during courtship while going hungry themselves. Birds that cache food inhibit immediate gratification for the sake of future need. There are many reasons to expect self-restraint in birds, therefore. The test results bear this out. Crows and ravens were given beans—a food they'd normally eat right away—after being taught that they could trade the beans later for a piece of sausage, which they liked better. The birds hung on to the beans for up to ten minutes.[29] When Griffin, the African gray of Irene Pepperberg, was tested on a similar paradigm, he managed even longer waiting times. The parrot had the advantage that he understood the instruction "Wait!" So while Griffin was sitting on his perch, a cup with a less preferred food, such as cereal, was put in front of him, and he was asked to wait. Griffin knew that if he waited long enough, he might get cashew nuts or even candies. If the cereal was still in the cup after a random time interval—anywhere from ten seconds to fifteen minutes—Griffin would receive the better food. He was successful 90 percent of the time, including on the longest delays.[30]

Most fascinating are the many ways in which children and animals cope with temptation. They are not passively sitting and staring at the object of desire but try to occupy themselves by creating distractions. Children avoid looking at the marshmallow, sometimes covering their eyes with their hands or putting their head into their arms. They talk to themselves, they sing, they invent games using their hands and feet, and they even fall asleep so as not to have to endure the terribly long wait.[31] The behavior of apes is not so different, and one study found that if given toys, apes are able to hold out longer. Toys help them take their attention off the candy machine. Or take Griffin, who about one-third through one of his longest waits threw the cup with cereal across the room. This way he didn't have to look at it. On other occasions, he moved the cup just out of reach, talked to himself, preened himself, shook his feathers, yawned extensively, or fell asleep (or at least closed his eyes). He also sometimes licked the treat without eating it, or shouted "Wanna nut!"

Some of these behaviors don't fit the situation at hand and fall under what ethologists call *displacement activities*, which occur when a drive is thwarted. This happens when two conflicting drives, such as fight and flight, arise at the same time. Since they cannot both be expressed, irrelevant behavior takes the pressure off. A fish spreading its fins to intimidate a rival may all of a sudden swim to the bottom to dig into the sand, or a rooster may interrupt a fight only to start pecking at some imaginary grains. In humans, a typical displacement activity is to scratch one's head when asked a tough question. Scratching is also common in other primates during cognitive tests, especially challenging ones.[32] Displacement activity occurs when motivational energy seeks an outlet and "sparks over" into extraneous behavior. The discoverer of this mechanism, the Dutch ethologist Adriaan Kortlandt, is still honored at the zoo in Amsterdam where he used to watch a colony of free-ranging cormorants. The wooden bench on which he spent hours following his birds is known as the "displacement bench." I recently sat on it and obviously couldn't resist yawning, fiddling, and scratching myself.

But this is not the whole explanation of how animals cope with delayed gratification, and why they preen themselves or yawn. There are cognitive interpretations, too. Long ago the father of American psychology, William James, proposed "will" and "ego strength" as the basis of self-control. This is how the behavior of children usually is interpreted, as in the following description of the marshmallow test: "The subject can wait most stoically if he expects that he really will get the deferred larger outcome in the waiting paradigm, and wants it very much, but shifts his attention elsewhere and occupies himself internally with cognitive distractions."[33] The emphasis here is on a deliberate, conscious strategy. The child knows what the future holds and wills his mind off the temptation in front of him. Given how similarly children and some animals behave under the same conditions, it is logical to favor the same explanation. Demonstrating impressive willpower, animals too may be aware of their own desires and try to curb them.

To explore this further, I visited Michael Beran, an American colleague at Georgia State University. Mike works at a lab in a large stretch of forest in Decatur, in the middle of the Atlanta area, with roomy accommodations for chimpanzees and monkeys. It is known as the Language Research Center, so named since Kanzi, the symbol-trained bonobo, was its first resident. At the same location, Charlie Menzel conducts tests of spatial memory on apes and Sarah Brosnan studies economic decision making by capuchins. The Atlanta area may well have the world's highest concentration of primatologists, since they are also found at Zoo Atlanta, in nearby Athens, Georgia, and of course at the Yerkes Primate Center, which historically sparked all this interest. As a result, we have expertise on a wide range of topics.

I asked Mike, who has worked extensively on self-control,[34] why articles in this field so often start out with the connection to consciousness, then quickly move to actual behavior without ever returning to the issue of consciousness. Are the authors teasing us? The reason, Mike felt, is that the link with consciousness is rather speculative. Strictly speaking, the fact that animals achieve a better outcome by waiting doesn't prove that they realize what will happen in the time ahead. On the other hand, their response doesn't depend on gradual learning, since they generally show it right away. This is why Mike regards self-control decisions as being future-oriented and cognitive. We may not have proof beyond all doubt, but the assumption is that the apes make these decisions based on the anticipation of a better outcome: "To argue that the behavior of apes is entirely under external stimulus control is silly to me."

Another argument for a cognitive interpretation is their behavior during long waits, which last up to twenty minutes, while candies drop at regular intervals into a bowl. The waiting apes like to play with things during this time, which suggests recognition that they need self-control. Mike described some of the weird things they do to keep themselves busy. Sherman (an adult male chimpanzee) would pick up a candy from the bowl, inspect it, then put it back. Or Panzee would disconnect the tube through which the candies roll in. She'd look at it and shake it before putting it back onto the dispenser. Given toys, they would use them as a distraction to make the wait easier. Such behavior hints at anticipation and strategizing, both of which suggest conscious awareness.

Mike's interest in this topic was triggered by a legendary experiment on reversal pointing by the American primatologist Sarah Boysen with Sheba, a chimpanzee. Sheba was asked to choose between two cups with different amounts of candy. The catch, however, was that the cup that she'd point at would go to another chimp, leaving her with the alternative cup. Obviously, the smart strategy would be for Sheba to reverse her pointing, indicating the cup with the *smaller* number of candies. Yet unable to overcome her desire for the fuller cup, she never learned to do so. When the candies were replaced by numbers, however, things changed. Sheba had learned the numbers 1 through 9, knowing the amounts of food associated with them. Presented with two different numbers, she never hesitated to point at the smaller one, showing that she understood how the reversals worked.[35]

Mike was impressed by Sally's research showing that chimps can't get the reversal right with actual candies. This was obviously a matter of self-control. When he tried the same test on his own chimps, they didn't pass either. Sally's idea to replace the candies with numbers was brilliant.

Whether it is the symbolizing or just the removal of the hedonic property, chimps trained with numerals were really good at it. When I asked if the same had ever been tried with children, Mike's answer reflected the deep concern of students of animal cognition with fair comparisons: "It may have been tried, I don't recall, but they probably explained it to the kids, and I would prefer that nothing is explained. We can't explain it to the apes either."

Know What You Know

The claim that only humans can mentally hop onto the time train, leaving all other species stranded on the platform, is tied to the fact that we consciously access past and future. Anything related to consciousness has been hard to accept in other species. But this reluctance is problematic: not because we know so much more about consciousness, but because we have growing evidence in other species for episodic memory, future planning, and delayed gratification. Either we abandon the idea that these capacities require consciousness, or we accept the possibility that animals may have it, too.

The fourth spoke on this wheel is *metacognition*, which is literally cognition about cognition, also known as "thinking about thinking." When the contestants in a game show are allowed to pick their topic, they obviously name the one they are most familiar with. This is metacognition in action, because it means they know what they know. In the same way, I may answer a question by saying "Wait, it's on the tip of my tongue!" In other words, I suspect that I know the answer, even though it's taking me time to recall it. A student raising her hand in class in reaction to a question is also relying on metacognition, because she only does so if she thinks she knows the solution. Metacognition rests on an executive function in the brain that allows one to monitor one's own memory. Again, we associate these processes with consciousness, which is exactly why metacognition, too, was deemed unique to our species.

Animal research in this area began perhaps with the *uncertainty response* noticed by Tolman in the 1920s. His rats seemed to hesitate before a difficult task as reflected in their "lookings or runnings back and forth."[36] This was most remarkable, since at the time animals were thought to simply respond to stimuli. Absent an inner life, why be in turmoil about a decision? Decades later the American psychologist David Smith gave a bottlenose dolphin the task to tell the difference between high and low tones. The dolphin was an eighteen-year-old male named Natua, in a pool at the Dolphin Research Center in Florida. As in Tolman's rats, Natua's level of confidence was quite manifest. He swam at different speeds toward the response, depending on how easy or hard it was to tell both tones apart. When they were very different, the dolphin arrived with such speed that his bow wave threatened to soak the electronics of the apparatus. They had to be covered with plastic. If the tones were similar, though, Natua slowed down, waggled his head, and wavered between the two paddles that he needed to touch in order to indicate a high or low sound. He didn't know which one to pick. Smith decided to make a study of Natua's uncertainty, mindful of Tolman's suggestion that it might reflect consciousness. The investigator created a way for the animal to opt out. A third paddle was added, which Natua could touch if he wanted a fresh trial with an easier distinction. The tougher the choice, the more Natua went for the third paddle, apparently realizing when he had trouble coming up with the right answer. Thus the field of animal metacognition was born.[37]

Investigators have essentially followed two approaches. One is to explore the uncertainty response, as in the dolphin study, while the other is to see if animals realize when they need more information. The first approach has been successful with rats and macaques. Robert Hampton, now a colleague at Emory University, gave monkeys a memory task on a touchscreen. They would first see one particular image, say a pink flower, then face a delay before being presented with several pictures, including the pink flower. The delay varied in length. Before each test,

A rhesus macaque knows that food has been hidden in one of four tubes, but he has no idea which one. He is not allowed to try every tube and will get only one pick. By bending down to first peek into the tubes, he demonstrates that he knows he doesn't know, which is a sign of metacognition.

the monkeys had the choice to either take it or decline it. If they took the test and correctly touched the pink flower, they gained a peanut. But if they declined, they only got a monkey pellet, a boring everyday food. The longer the delay, the more the monkeys declined taking the test despite its better reward. They seemed to realize that their memory had faded. Occasionally, they were forced to take a trial without a chance of escape. In those cases they fared rather poorly. In other words, they opted out for a reason, doing so when they couldn't count on their memory.[38] A similar test with rats gave similar results: the rats performed best on tests that they had deliberately chosen to take.[39] In other words, both macaques and rats volunteer for tests only when they feel confident, suggesting that they know their own knowledge.

The second approach concerns information seeking. For example, jays placed at peepholes were given an opportunity to watch food—waxworms—being hidden before they were allowed to enter the area to find it. They could look through one peephole to see an experimenter put a waxworm in one of four open cups, or they could look through another to see another experimenter with three covered cups plus one open one. In the second case, it was obvious where the worm would end up. Before entering the area to find the worm, the birds spent more time watching the first experimenter. They seemed to realize that this was the information they needed most.[40]

In monkeys and apes, the same sort of test has been done by having them watch an experimenter hide food in one of several horizontal pipes. Obviously, the primates remembered where he had put the food and confidently selected the correct pipe. If the food hiding had taken place in secret, however, they were not sure which pipe to pick. They peeked into the pipes, bending down to get a good look, before selecting one. They realized that they needed more information to succeed.[41]

As a result of these studies, some animals are now believed to track their own knowledge and to realize when it is deficient. It all fits Tolman's insistence that animals are active processors of the cues around them, with beliefs, expectations, perhaps even consciousness. This viewpoint being on the rise, I asked my colleague Rob Hampton about the state of affairs in this field. The two of us have offices on the same floor of Emory's psychology department. While sitting in mine, we first watched the video of Lisala carrying her huge rock. Like a real scientist, Rob immediately began to imagine how to turn this situation into a controlled experiment by varying the locations of the nuts and the tools, even though for me the beauty of the whole sequence was Lisala's spontaneity. We had nothing to do with it. Rob was impressed.

I asked him if his work on metacognition had been inspired by the dolphin study, but he rather saw this as a case of convergent interests. The dolphin study did come out first, but it wasn't about memory, which was Rob's focus. He was inspired by the ideas of Alastair Inman, a postdoc in Sara Shettleworth's Toronto lab, where Rob worked at the time. Alastair wondered about the cost of memorizing things. What is the price of holding information in mind? He set up an experiment on pigeon memory that was similar to the metacognition test for monkeys that Rob developed.[42]

When I asked what he thought of people who draw a sharp line between humans and other animals, such as Endel Tulving's shifting definitions, Rob exclaimed: "Tulving! He loves to do that. He has done a great service to the animal research community." Tulving says those things, Rob believes, because he thinks it's fun to set a high bar. He knows that others will go after it, so he pushes them to come up with clever experiments. In his first monkey paper, Rob thanked Tulving for his "incitement." Meeting the senior scientist not long thereafter at a conference, Tulving told Rob, "I have seen what you wrote, thank you!"

For Rob, the big question in relation to consciousness is why we actually need it. What is it good for? After all, there are lots of things we can do unconsciously. For example, amnesic patients are able to learn without knowing what they have learned. They may learn to make inverse drawings guided by a mirror. They acquire the hand-eye coordination at about the same rate as any other person, but every time you test them, they'll tell you that they've never done it before. It is all new to them. In their behavior, though, it is obvious that they have experience with the task and have acquired the required skill.

While consciousness has evolved at least once, it is unclear why and under which conditions. Rob considers it such a messy word that he is reluctant to use it. He adds, "Anyone who thinks they have solved the problem of consciousness hasn't been thinking about it carefully enough."

Consciousness

When in 2012 a group of prominent scientists came out with *The Cambridge Declaration on Consciousness*, I was skeptical.[43] The media described it as asserting once and for all that nonhuman animals are conscious beings. Like most scientists studying animal behavior, I really don't know what to say to this. Given how ill-defined consciousness is, it is not something we can affirm by majority vote or by people saying "Of course, they are conscious—I can see it in their eyes." Subjective feelings won't get us there. Science goes by hard evidence.

But in reading the actual declaration, I calmed down, because it is a reasonable document. It doesn't actually claim animal consciousness, whatever that is. It only says that given the similarities in behavior and nervous systems between humans and other large-brained species, there is no reason to cling to the notion that only humans are conscious. As the document puts it, "The weight of evidence indicates that humans are not unique in possessing the neurological substrates that generate consciousness." I can live with that. As you can see from this chapter, there is sound evidence that mental processes associated with consciousness in humans, such as how we relate to the past and future, occur in other species as well. Strictly speaking, this doesn't prove consciousness, but science is increasingly favoring continuity over discontinuity. This is certainly true for comparisons between humans and other primates, but extends to other mammals and birds, especially since bird brains turn out to resemble those of mammals more than previously thought. All vertebrate brains are homologous.

Although we cannot directly measure consciousness, other species show evidence of having precisely those capacities traditionally viewed as its indicators. To maintain that they possess these capacities in the absence of consciousness introduces an unnecessary dichotomy. It suggests that they do what we do but in fundamentally different ways. From an evolutionary standpoint, this sounds illogical. And logic is one of those other capacities we pride ourselves on.

8 | OF MIRRORS AND JARS

Pepsi was the star of a recent study on Asian elephants. The adolescent bull passed a mirror test conducted by Joshua Plotnik by carefully touching a large white X that had been painted on the left-hand side of his forehead. He never paid attention to the X that had been put with invisible paint on the right-hand side; nor did he touch the white one until he walked up to the mirror in the middle of a meadow. The next day we reversed the sides of the visible and invisible markings, and Pepsi again specifically felt the white X. He rubbed off some of the paint with the tip of his trunk and brought it to his mouth, tasting it. Since he could know its location only via his reflection, he must have connected his mirror image with himself. As if to make the point that the mark test isn't the only way to do so, Pepsi took one step back at the end of testing to open his mouth wide. With the mirror's help, he peered deeply inside. This move, also common in apes, makes perfect sense given that one never gets to see one's own tongue and teeth without a mirror.[1]

Years later Pepsi towered over me as a nearly adult male. He was very gentle, though, lifting me up and putting me down on the orders of his mahout. Revisiting Thailand to see the camp in the Golden Triangle, where the Think Elephants International Foundation conducts its research, I met Josh's team of enthusiastic young assistants. Every day they invite a couple of elephants to their experiments. With a mahout

A marked Asian elephant in front of a mirror. The mark test requires an individual to connect her reflection with her own body, resulting in inspection of the mark. Only a handful of species pass this test spontaneously.

sitting high up on their neck, the colossal animals lumber to the testing site on the jungle's edge. After the mahout gets off to squat in the background, the elephant performs a few simple tasks. She feels an object with her trunk, after which she is asked to pick a matching one from among several; or she stretches her trunk to smell the difference between two buckets depending on what the students put into them.[2]

Everyone knows that elephants are smart, but there is an enormous scarcity of data similar to those for primates, corvids, dogs, rats, dolphins, and so on. All we have for the elephant is spontaneous behavior, which doesn't allow for the precision and controls that science desires. Discrimination tasks like the ones I witnessed are an excellent starting point. But even if the pachyderm mind may be the next frontier in evolutionary cognition, it is a most challenging one given that the elephant is probably the only land animal never to be seen alive on a university campus or in a conventional lab. While science's preference for easy-to-keep species is understandable, it has its limits. It has given us a small-brain perspective on animal cognition, one that we have had trouble shedding.

Elephants Listening

Southeast Asians have a long-standing cultural relation with elephants. For thousands of years, these animals have carried out heavy forest work, transported royalty, and served in hunting and warfare. They have always remained wild, though. The species is not domesticated in the genetic sense, and free-ranging elephants still often sire the offspring of captive ones. Not surprisingly, elephants are less predictable than many domesticated animals. They can be hostile to people, occasionally killing a mahout or tourist, but many of them also form lifelong bonds with their caretakers. In one story, an elephant at the age of ten pulled her drowning mahout out of a lake after hearing his cries for help a kilometer away; in another, a fully grown bull would charge anyone who came close except the wife of the village elder, whom he would caress with his trunk. Young elephants grow so used to people that they learn how to fool them by stuffing a trunkful of grass into the wooden bells around their necks so as to muffle their sound. This way they can move about unnoticed.

African elephants, in contrast, are rarely brought under human control. They live their own parallel lives, even though the massive ivory trade is now putting them in danger to the point that we face the dismaying prospect of permanently losing one of the world's most beloved and charismatic animals. The elephant's *Umwelt* being largely acoustic and olfactory, the protection of wild populations against poaching and conflict with humans requires methods that are not immediately obvious to our visual species. Studies focus on the extraordinary senses of these animals. One study, in arid Namibia, followed free-ranging elephants equipped with GPS collars. It discovered that these animals are aware of thunderstorms at enormous distances and adjust their travel routes to precipitation days before it actually arrives. How do they do this? Elephants can hear infrasound, which are sound waves far below the human hearing range. Also used in communication, these sounds travel over much longer distances than the ones we are able to discern.[3] Is it possible that elephants can hear thunder and rainfall hundreds of miles away? It seems the only way to explain their behavior.

But isn't this just a matter of perception? Cognition and perception cannot be separated, though. They go hand in hand. As the father of cognitive psychology, Ulric Neisser, put it: "the world of experience is produced by the man who experiences it."[4] Since the late Neisser was a colleague of mine, I know that nonhuman minds were not his foremost interest, yet he refused to view animals as mere learning machines. The behaviorist program was ill suited to all species, he felt, not just ours. Instead, he emphasized perception and how it is turned into experience by picking and choosing what sensory input to pay attention to and how to process and organize it. Reality is a mental construct. This is what makes the elephant, the bat, the dolphin, the octopus, and the star-nosed mole so intriguing. They have senses that we either don't have, or that we have in a much less developed form, making the way they relate to their environment impossible for us to fathom. They construct their own realities. We may attach less significance to these, simply because they are so alien, but they are obviously all-important to these animals. Even when they process information familiar to us, they may do so quite differently, such as when elephants tell human languages apart. This ability was first demonstrated in African elephants.

In Amboseli National Park, in Kenya, the British ethologist Karen McComb studied elephant reactions to different human ethnic groups. The cattle-herding Maasai sometimes spear elephants in order to show their virility or to gain access to grazing grounds and water holes. Understandably, elephants flee the Maasai, who approach them in their characteristic red ochre robes, but they don't avoid other people on foot.[5] How do they recognize the Maasai? Instead of focusing on their color vision, McComb explored what is perhaps the elephant's keenest sense: sound. She contrasted the Maasai with the Kamba people, who live in the same area but rarely interfere with elephants. From a concealed loudspeaker, McComb played human voices saying a single phrase in the language of either the Maasai or the Kamba: "Look, look over there, a group of elephants is coming." It is hard to imagine that the precise words mattered, but the investigators compared the elephants' reaction to the voices of adult men, adult women, and boys.

Herds retreated and "bunched" together (forming a tight circle with calves in the middle) more often after playbacks of Maasai than of Kamba voices. Maasai male voices triggered more defensive reactions than those of Maasai women and boys. Even after the natural voices were acoustically transformed so as to make male voices sound more female, and vice versa, the outcome remained the same. The elephants were especially vigilant upon hearing the resynthesized voices of Maasai men. This was surprising because the pitch of these voices had now the opposite gender's qualities. Possibly the elephants identified gender by other characteristics, such as the fact that female voices tend to be more melodious and "breathy" than those of males.[6]

Experience played a role, because herds led by older matriarchs were more discriminating. The same difference was found in another study in which lion roaring was played from a speaker. Older matriarchs would charge the speaker, which is quite different from their hasty retreat from Maasai voices.[7] Aggressive mobbing of men carrying spears is unlikely to pay off, yet driving off lions is something elephants are good at. Despite their size, these animals face other dangers, including very small ones, such as stinging bees. Elephants are vulnerable to stings around the eyes and up their trunks, and young elephants lack a thick enough skin to protect themselves against a mass attack. Elephants give deep rumbles as an alarm to both humans and bees, but the two sounds must differ because playbacks induce quite different responses. Upon hearing the bee-rumble from a speaker, for example, elephants flee with head-shaking movements that would knock insects away, a reaction not shown to the human-rumble.[8]

In short, elephants make sophisticated distinctions regarding potential enemies to the point that they classify our own species based on language, age, and gender. How they do so is not entirely clear, but studies like these are beginning to scratch the surface of one of the most enigmatic minds on the planet.

The Magpie in the Mirror

The ability to recognize oneself in the mirror is often viewed in absolute terms. According to Gallup, the pioneer of this field, a species either passes the mirror mark test and is self-aware, or it doesn't and isn't.[9] Very few species do. For the longest time, only humans and the great apes passed, and not even all those. Gorillas used to flunk the mark test, leading to theories about why the poor things might have lost their self-awareness.[10]

Evolutionary science, however, is uncomfortable with black-and-white distinctions. It is hard to imagine that among any set of related species, some are self-aware whereas others, for lack of a better term, remain unaware. Every animal needs to set its body apart from its surroundings and to have a sense of agency (awareness that it controls its own actions).[11] You wouldn't want to be a monkey up in a tree without awareness of how your own body will impact a lower branch on which you intend to land. And you wouldn't want to engage in rough-and-tumble play with a fellow monkey, with all your combined arms, legs, and tails intertwined, while stupidly gnawing on your own foot or tail! Monkeys never make this mistake and gnaw exclusively on their part-

ner's foot or tail in such a tangle. They have a well-developed body own-
ership and self-other distinction.

In fact, experiments on the sense of agency show that species without mirror self-recognition are very well capable of distinguishing their
own actions from those produced by others. Tested in front of a computer screen, they have no trouble telling the difference between a cursor
that they themselves control with a joystick and a cursor that moves by
itself.[12] Self-agency is part of every action that an animal—any animal—
undertakes. In addition, some species may possess their own unusual
kind of self-recognition, such as bats and dolphins that pick out the echoes
of their own vocalizations from among the sounds made by others.

Cognitive psychology doesn't like absolute differences either, but for
a different reason. The problem with the mirror test was that it introduced the *wrong* absolute difference. Instead of sharply dividing humans
from all other animals—which, as we have seen, is a staple of the field—
Gallup's mirror test moved the Rubicon slightly to annex a few more
species. Lumping humans in with the apes so as to elevate the Hominoids, as a group, to a different mental level than the rest of the animal
kingdom, didn't go over well. It diluted humanity's special status. Still
today, claims about self-awareness outside our own species cause con-
sternation, and debates about mirror responses turn acrimonious. Moreover, many specialists have felt the need to conduct mirror tests on the
animals in their care, usually with disappointing results. These debates
have led me to the sarcastic conclusion that mirror self-recognition is
considered a big deal only by scientists working on the handful of species capable of it, whereas all others poo-poo the phenomenon.

Since I study animals that both do and don't recognize themselves in
the mirror, and have a high opinion of them all, I feel torn. I do think that
spontaneous self-recognition means something. It may signal a stronger
self-identity, such as is also reflected in perspective taking and targeted
helping. These capacities are most marked in animals that pass the mir-

ror test as well as in children who have reached the age, around two, when they do. This is also the age when they can't stop referring to themselves, as in "Mama, look at me!" Their sharpened self-other distinction is said to help them adopt another's viewpoint.[13] Still, I can't believe that a sense of self is absent either in other species or in younger children. Rather obviously, animals that fail to link their mirror image to their own body vary greatly in what they understand. Small songbirds and fighting fish, for example, never get over their mirror image and keep courting or attacking it. During the spring, when they are most territorial, tits and bluebirds will respond this way to the sideview mirror of a car and stop their hostilities only when the car drives off. This is absolutely not what monkeys do, nor many other animals. We would not be able to have mirrors in our homes if cats and dogs reacted the same way. These animals may not recognize themselves, but they are also not totally baffled by the mirror, at least not for long. They learn to ignore their reflection.

Some species go further in that they understand mirror basics. Monkeys, for example, may not recognize themselves but are able to use the mirror as a tool. If you hide food that can be found only by using a mirror to look around a corner, the monkey will have no trouble reaching for it. Many a dog can do the same: holding up a cookie behind them while they watch you in a mirror makes them turn around. Curiously, it is specifically the relation with their own body, their own self in the mirror, that they fail to grasp. But even then, rhesus monkeys can be taught to do so. It requires adding a physical sensation. They need a mark that they can both see in the mirror and feel on their body, such as a laser light that irritates the skin or a cap fastened to their head. Instead of a traditional mark test, this is better described as a *felt* mark test. Only under these circumstances can monkeys learn to connect their reflection with their own body.[14] This is obviously not the same as what apes do spontaneously relying on vision alone, but it does suggest that some of the underlying cognition is shared.

Even though capuchin monkeys fail the visual mark test, we decided to study them in a way that, surprisingly, no one had ever tried before. Our goal was to see if these monkeys truly mistake their reflection for a "stranger," as is commonly implied. Capuchins were placed in front of a Plexiglas panel, behind which they faced either a member of their own group, a stranger of their species, or a mirror. It quickly became evident that the mirror was special. They treated their reflection quite differently from a real monkey. They didn't need any time to decide what they saw, and reacted within seconds. They turned their backs to strangers, barely glancing at them, yet made prolonged eye contact with their own reflections as if thrilled to see themselves. They showed absolutely none of the timidity toward the mirror image that one would expect if they mistook it for a stranger. Mothers, for example, let their infants freely play in front of the mirror yet held them close in case of a stranger. But the monkeys also never inspected themselves in the mirror, the way apes do all the time, or the way Pepsi the elephant had done. They never opened their mouth to peek inside. Thus, while capuchins fail to recognize themselves, they also don't mix up their reflection with someone else.

As a result, I have become a gradualist.[15] There are many stages of mirror understanding, running all the way from utter confusion to a full appreciation of the specular image. These stages are also recognizable in human infants, which are curious about their mirror image well before passing the mark test. Self-awareness develops like an onion, building layer upon layer, rather than appearing out of the blue at a given age.[16] For this reason, we should stop looking at the mark test as the litmus test of self-awareness. It is only one of many ways to find out about the conscious self.

Nevertheless, it remains fascinating how few species pass this test without a helping hand. After the Hominoids, spontaneous self-recognition was observed only in elephants and dolphins. When bottlenose dolphins at the New York Aquarium were marked by Diana Reiss and Lori Marino with painted dots, they would race from the spot where the marking took place to a mirror in another pool, at quite a distance, only to spin around seemingly to get a good look at themselves. The dolphins spent more time near the mirror checking out their bodies when they had been marked than without a visible mark.[17]

It was unavoidable that the mirror test would be tried on birds. While most species have thus far failed, we have one exception: the Eurasian magpie. It is an interesting species to put in front of a reflective surface. As a child, I learned never to leave small shiny objects, such as teaspoons, unattended outdoors as these raucous birds will steal anything they can put their beaks on. This folklore inspired a Rossini opera, *La gazza ladra* (*The Thieving Magpie*). Nowadays, this view has been replaced with a more ecologically sensitive one that depicts magpies as murderous robbers of the nests of innocent songbirds. Either way, they are considered black-and-white gangsters.

But no one has ever accused a magpie of being stupid. The bird belongs to the corvid family that has begun to challenge the cognitive supremacy of primates. The German psychologist Helmut Prior subjected magpies to a mirror test that was at least as well controlled as any conducted on apes and children. Placed on their black bib (throat feathers), the mark—a tiny yellow sticker—stood out but was visible only with help of the mirror. The birds were untrained, which is a critical difference with the highly coached pigeons employed long ago to discredit mirror research. Put in front of a mirror, the magpies kept scratching with their foot until the mark was gone. They never did the same amount of frantic scratching if there was no mirror to see themselves in, and they ignored a "sham" mark—a black sticker on their black bib. As a result, the self-recognition elite has now been expanded with its first feathered member. Others may follow.[18]

The next frontier will be to see if animals *care* about their mirror image to the point of embellishing themselves, the way we do with makeup, hair care, earrings, and the like. Does the mirror induce vanity? Would any species other than ours be prone to take selfies, if they

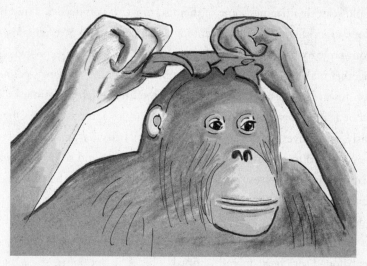

Suma, an orangutan at a German zoo, loved to decorate herself in front of the mirror.
Here she puts a leaf of lettuce onto her head like a hat.

could? This possibility was first hinted at by observations in the 1970s
of a female orangutan at the Osnabrück Zoo, in Germany. Jürgen Lethmate and Gerti Dücker described Suma's narcissistic ways:

> She gathered salad and cabbage leaves, shook each leaf and piled them up.
> Eventually, she placed one leaf on her head and walked straight to the mir-
> ror with it. She sat down directly in front of it, contemplated her headcover
> in the mirror, straightened it a bit with her hand, squashed it with a fist,
> then put the leaf on her forehead and began to bob up and down. Later,
> Suma arrived holding a salad leaf in her hand at the bars [where the mirror
> stood] to lay it on her head once she could see herself in the mirror.[19]

The Mollusk Mind

As a biology student, my favorite textbook was *Animals Without Backbones*. It may seem an odd choice given my current interests, but I was
awestruck by all the exotic life-forms that I had never heard about or

could scarcely imagine, some of them so tiny that you needed a microscope to see them. The book went into great detail of all invertebrates—from protozoans and sponges to worms, mollusks, and insects—which together make up 97 percent of the animal kingdom.[20] Whereas cognition research focuses almost entirely on the tiny vertebrate minority, it is not as if the rest doesn't move, eat, mate, fight, and cooperate. Obviously, some invertebrates show more complex behavior than others, but they all need to pay attention to their surroundings and solve problems that present themselves. In the same way that almost all these animals have reproductive organs and digestive tracts, they can't survive without a degree of cognition.

The brainiest of the bunch is the octopus, which is a soft-bodied cephalopod, or "head-footed" animal. This is an apt name, since their squishy bodies consist of a head that directly joins eight limbs, while the body (the mantle) is positioned behind the head. The cephalopods are an ancient class that arose well before there were land vertebrates around, but the group to which the octopus belongs is a fairly modern

The octopus has a most remarkable nervous system that allows it to solve challenging problems, such as how to escape from a glass jar closed with a screw top.

offshoot. We seem to have almost nothing in common with them, both anatomically and mentally. Yet they have been reported to open a pill bottle protected by a childproof cap. Since this requires the cap to be pushed down and twisted at the same time, it takes skill, intelligence, and persistence. Some public aquariums show off octopus intelligence by locking the animal in a glass jar that they close with a screw top. Like a true Houdini, the octopus takes less than a minute to grab the cover from within with its suckers and unscrew it so as to escape.

However, when octopuses were given a transparent jar that contained a live crayfish, they failed to do anything. This greatly puzzled the scientists, because the delicacy was clearly visible and moving about. Do octopuses perhaps have trouble unscrewing a lid from the outside? It turned out to be one of those human misjudgments. Despite having excellent eyes, octopuses rarely rely on vision to catch prey. They use mainly touch and chemical information and fail to recognize prey without those cues. As soon as the jar was smeared on the outside with herring "slime," making it taste like fish, the octopus swung into action and started manipulating it until the top came off. It quickly removed the crayfish and ate it. With further skill development, the process became routine.[21]

In captivity, octopuses react to us in ways that we find hard not to anthropomorphize. One octopus was fond of raw chicken eggs—each day it would accept an egg and break it to suck out its contents. One day, however, this octopus accidentally received a rotten egg. Upon noticing, it shot the egg's smelly remains over the edge of its tank back at the surprised human from whom it had received it.[22] Given how well they distinguish people, octopuses probably remember encounters like these. In a recognition test, an octopus was exposed to two different persons, one of whom consistently fed it, whereas the other mildly poked it with a bristle on a stick. Initially, the animal made no distinction, but after several days it began doing so despite the fact that both humans wore identical blue overalls. Seeing the loathsome person, the octopus would withdraw, emit jets of water with its funnel, and show a dark bar through its eyes—a color change associated with threat and irritation. It would approach the nice person, on the other hand, without making any attempt at drenching her.[23]

The octopus brain is the largest and most complex of all invertebrates, but the explanation of its extraordinary skills may lie elsewhere. These animals literally think outside the box. Each octopus has nearly two thousand suckers, every single one equipped with its own ganglion with half a million neurons. That amounts to a lot of neurons on top of a 65-million-neuron brain. In addition, it has a chain of ganglia along its arms. The brain connects with all these "mini brains," which are also joined among themselves. Instead of a single central command, as in our species, the cephalopod nervous system is more like the Internet: there is extensive local control. A severed arm may crawl on its own and even pick up food. Similarly, a shrimp or small crab can be handed from one sucker to the next, as if on a conveyer belt, in the direction of the octopus's mouth. When these animals change skin color in self-defense, the decision may come from central command, but perhaps the skin is involved as well, since cephalopod skin may detect light. It sounds rather unbelievable: an organism with seeing skin and eight independently thinking arms![24]

This realization has led to a bit of hype: that the octopus is the most intelligent organism in the ocean, a sentient being that we should stop eating. We shouldn't overlook dolphins and orcas, though, which have vastly larger brains. Even if the octopus stands out among invertebrates, its tool use is rather limited, and its reaction to a mirror is as perplexed as that of a small songbird. It remains unclear whether an octopus is smarter than most fish, but let me hasten to add that such comparisons barely make any sense. Instead of turning the study of cognition into a contest, we should avoid putting apples next to oranges. The octopus's senses and anatomy, including its decentralized nervous system, make it unparalleled.

If superlatives of uniqueness were allowed, the octopus might be the most unique species of them all. They defy comparison with any other group, unlike our own species, which derives from a long line of land vertebrates with structurally similar body plans and brains.

Octopuses have an odd life cycle. Most live only one or two years, which is unusual for an animal with their brainpower. They grow fast while trying to stay away from predators until they have a chance to mate and reproduce, after which they die. They stop eating, lose weight, and go into senescence.[25] This is the stage about which Aristotle observed: "after giving birth . . . [they] become stupid, and are not aware of being tossed about in the water, but it is easy to dive and catch them by hand."[26]

These short-lived loners have no social organization to speak of. Given their biology, they have no reason to pay attention to one another, except as rivals, mates, predators, and prey. They are certainly not friends or partners. There is no evidence that they learn from others or spread behavioral traditions, the way many vertebrates, including fish, do. The absence of social bonds and cooperation, and their cannibalistic ways, make cephalopods quite alien to us.

Their main worry is predation, because apart from their own kind, they are eaten by almost everything around, from marine mammals, diving birds, sharks, and other fish to humans. When they get larger, they become formidable predators themselves, as the Seattle Aquarium accidentally found out. Worried about their giant Pacific octopus in a tank full of sharks, staff were hoping that the animal would know how to hide. But then they noticed one dogfish (a small shark) after another disappearing from the tank—and found to their astonishment that the octopus had turned the tables. The octopus may be the only playful invertebrate. I say *may* since play behavior is almost impossible to define, but the octopus appears to go beyond mere manipulation and checking out of novel objects. The Canadian biologist Jennifer Mather found that given a new toy, the animal will move from exploration ("What is this?") to repeated lively movements and tossing around ("What can I do with it?"). With their funnel, they blow jets of water at a floating plastic bottle, for example, to move it from one side of their tank to another, or to have it tossed back at them by the water flow of the filter, which makes them look as if they were bouncing a ball. Such manipulations, which serve no obvious purpose and are repeated over and over, have been taken as indications of play.[27]

Tied to the immense predation pressure under which these animals live is their ability for camouflage. Perhaps their most astonishing specialization, it provides an inexhaustible "magic well" for those who study them.. The octopus changes color so rapidly that it out-chameleons the chameleon. Roger Hanlon, a scientist at the Marine Biological Laboratory in Woods Hole, Massachusetts, has collected rare underwater footage of octopuses in action. All we see at first is a clump of algae on a rock, but hidden among it is a large octopus indistinguishable from its surroundings. When the approaching human diver scares the animal, it turns almost white, revealing that it represented almost half the clump of algae. It speeds away while shooting a dark cloud of ink, which is its secondary defense. The animal then lands on the sea floor and makes itself look huge by spreading all its arms and stretching the skin between them into a tent. This frightening expansion is its tertiary defense.

When this video clip is slowed down and played backward, it is easy to see how superb the original camouflage was. Both structurally and color-wise, the large octopus had made itself look exactly like an algae-covered rock. It did so by making its chromatophores (millions of neurally controlled pigment sacs in its skin) match their surroundings. But instead of exactly mimicking its background, which is impossible, it did so just well enough to fool our visual system. And it probably did more than that, since the octopus also takes other visual systems into account. Humans see no polarized or ultraviolet light and don't have great night vision, whereas the octopus's camouflage needs to trick all these visual capacities. In doing so, it draws on a limited set of patterns that it has in stand-by mode. Turning on one of these "blueprint" patterns allows it to blend in in a fraction of a second. The result is an optical illusion, but one realistic enough to save its life hundreds of times.[28]

Sometimes an octopus mimics an inanimate object, such as a rock or plant, while moving so slowly that one would swear it is not moving at all. It does so when it needs to cross an open space, an activity that exposes it to detection. Imitating a plant, the octopus waves some of its arms above itself, making them look like branches, while tiptoeing on three or four of its remaining arms. It takes tiny little steps in line with the water movements. If the ocean is wild, plants sway back and forth, which helps the octopus disguise its steps by swaying in the same rhythm. On a waveless day, on the other hand, nothing else moves, so the octopus needs to be extremely careful. It may take twenty minutes to cross a stretch of sea floor that it otherwise might have crossed in twenty seconds. The animal acts as if rooted to the spot, counting on the fact that no predator will take the time to notice that it is actually inching forward.[29]

The champion of camouflage, finally, is the mimic octopus, a species found off the coast of Indonesia that impersonates other species. It acts like a flounder by adopting this fish's body shape and color as well as its typical undulating swimming pattern close to the sea floor. The repertoire of this octopus includes adopting the likeness of a dozen local marine organisms, such as lionfish, sea snakes, and jellyfish.

We don't know exactly how octopuses achieve this astonishing range of mimicry. Some of it may be automated, but there is probably also learning involved based on observations of other creatures and adoption of their habits. As primates, we find it impossible to relate to these remarkable capacities, and we may hesitate to call them cognitive. We tend to view invertebrates as instinct machines, arriving at solutions through inborn behavior. But this position has become untenable. There are too many remarkable observations—including the deceptive tactics of cuttlefish, close relatives of the octopus.

Male cuttlefish courting a female may trick rival males into thinking there is nothing to worry about. The courting male adopts the coloring of a female on the side of his body that faces his rival, so that the latter believes he is looking at a female. But the same male keeps his original coloring on the female's side of his body in order to keep her interested. He thus courts her surreptitiously. This two-faced tactic, called dual-gender signaling, suggests tactical skills of an order that we might expect in primates but not mollusks.[30] Hanlon rightly claims that cephalopod truth is stranger than fiction.

Invertebrates will probably continue to challenge students of evolutionary cognition. Being anatomically quite different yet facing many of the same survival problems as the vertebrates, they offer fertile grounds for convergent cognitive evolution. Among the arthropods, for example, we find jumping spiders known to trick other spiders into thinking that their web contains a struggling insect. When the web-owner hurries over for the kill, she herself becomes the prey. Instead of knowing at birth how to enact a trapped insect, jumping spiders seem to learn how to do so by trial and error. They try out a kaleidoscope of random pluckings and vibrations on the silk of another spider, using their palps and legs, while taking note which signals best lure the owner toward them. The most effective signals will be repeated on future occasions. This tactic allows them to fine-tune their mimicry to any victim species, which is why arachnologists have begun to speak of spider cognition.[31]

And why not?

When in Rome

To our surprise, chimps turn out to be conformists. Copying others for one's own benefit is one thing, but wanting to act like everybody else is quite another. It is the foundation of human culture. We discovered this tendency when Vicky Horner presented two separate groups of chimpanzees with an apparatus from which food could be extracted in two different ways. The apes could either poke a stick into a hole to release a grape or use the same stick to lift up a little trap and a grape would roll out. They learned the technique from a model: a pretrained group member. One group saw a lifting model, the other a poking model. Even though we used the same apparatus for both groups, moving it back and forth between them, the first learned to lift, and the second to poke. Vicky had created two distinct cultures, dubbed the "lifters" and the "pokers."[32]

There were exceptions, though. A few individuals discovered both techniques or used a different one than their model had demonstrated. When we retested the chimps two months later, though, most of the exceptions had vanished. It was as if all the apes had settled on a group norm, following the rule "Do what everyone else is doing regardless of what you found out by yourself." Since we never noticed any peer pressure nor any advantage of one technique over the other, we attributed this uniformity to a *conformist bias*. Such a bias obviously fits my ideas about imitation guided by a sense of belonging as well as what we know about human behavior. Members of our own species are the ultimate conformists, going so far as abandoning their personal beliefs if they collide with the majority view. Our openness to suggestion goes well beyond what we found in the chimps, yet it seems related. This is why the conformist label stuck.[33]

It is increasingly applied to primate culture, such as by Susan Perry in her fieldwork on capuchin monkeys. Perry's monkeys have two equally efficient ways of shaking the seeds out of the *Luehea* fruits that they encounter in the Costa Rican jungle. They can either pound the fruits or rub them on a tree branch. Capuchins are the most vigorous and enthusiastic foragers I know, and most adults develop one technique or the other but not both. Perry found conformism in daughters, who adopted the preferred method of their mothers, but not in sons.[34] This sex difference, also known of juvenile chimpanzees learning to fish for termites with twigs, makes sense if social learning is driven by identification with the model. Mothers act as role models for daughters but not necessarily for sons.[35]

Conformism is hard to substantiate in the field. There are too many alternative explanations for why one individual might act like another, including genetic and ecological ones. How these issues can be resolved was shown by a large-scale project on humpback whales in the Gulf of Maine in the northeastern United States. In addition to their regular bubble-feeding, in which whales drive fish together with air bubbles, one male invented a new technique. First seen in 1980, this whale would whack the ocean surface with his fluke to produce a loud noise that clumped the prey even more. Over time this lobtail technique became increasingly common in the population. In the course of a quarter-century, investigators carefully plotted how it spread across six hundred individually recognized whales. They found that whales who had associated with those employing the technique were more likely to use it themselves. Kinship could be ruled out as a factor, because whether a whale had a lobtail-feeding mother hardly mattered. It all boiled down to whom they had encountered while feeding on fish. Since large cetaceans are unsuitable for experiments, this may be as close as we will ever get to proving that a habit spread socially as opposed to genetically.[36]

On wild primates, experimental work is rare for different reasons. First of all, these animals are neophobic, and rightly so, because imagine the danger of freely approaching human contraptions, including those set by poachers. Second, fieldworkers generally hate to expose their animals to artificial situations, since their goal is to study them with as little disturbance as possible. Third, they have no control over who participates in an experiment and for how long, thus precluding the kind of tests typically applied to animals in captivity.

So one has to admire one of the most elegant experiments on conformism on wild monkeys, carried out by the Dutch primatologist Erica van de Waal (no relation).[37] Teaming up with Andy Whiten, who has been an engine of cultural studies, van de Waal gave vervet monkeys in a South African game reserve open plastic boxes filled with maize corn. These small grayish monkeys with black faces love corn, but there was a catch: the scientists had manipulated the supply. There were always two boxes with two colors of corn, blue and pink. One color was good to eat whereas the other was laced with aloe, making it disgusting. Depending on which color corn was palatable, and which not, some groups learned to eat blue, and others pink.

This preference is easily explained by associative learning. But then the investigators removed the distasteful treatment and waited for infants to be born and new males to immigrate from neighboring areas. They watched several groups of monkeys that were supplied with perfectly fine corn of both colors. All adults stubbornly stuck to their acquired preference, however, and never discovered the improved taste of the alternative color. Twenty-six of twenty-seven newborn infants learned to eat only the locally preferred food. Like their mothers, they didn't touch the other color, even though it was freely available and just as good as the other. Individual exploration was obviously suppressed. The youngsters might even sit on top of the box with the rejected corn while happily feeding on the other type. The single exception was an infant whose mother was so low in rank, and so hungry, that she occasionally tasted the forbidden fruits. Thus, all newborns copied their mothers' feeding habits. Male immigrants, too, ended up adopting the local color even if they arrived from groups with the opposite preference. That they switched their preference strongly suggests conformism, since these males knew from experience that the other color was perfectly edible. They simply followed the adage "When in Rome . . ."

These studies prove the immense power of imitation and conformism. It is not a mere extravagance that animals occasionally engage in for trivial reasons—which, I hate to say, is how animal traditions have sometimes been derided—but a widespread practice with great survival value. Infants who follow their mother's example of what to eat and what to avoid obviously stand a better chance in life than infants who try to figure out everything on their own. The idea of conformism among animals is increasingly supported for social behavior as well. One study tested both children and chimpanzees on generosity. The goal was to see if they were prepared to do a member of their own species a favor at no cost to themselves. They indeed did so, and their willingness increased if they themselves had received generosity from others—*any* others, not just their testing partner. Is kind behavior contagious? Love begets love, we say, or as the investigators put it more dryly, primates tend to adopt the most commonly perceived responses in the population.[38]

The same can be concluded from an experiment in which we mixed two different macaques: rhesus and stumptail monkeys. Juveniles of both species were placed together, day and night, for five months. These macaques have strikingly different temperaments: rhesus are a quarrelsome, nonconciliatory bunch, whereas stumptails are laid-back and pacific. I sometimes jokingly call them the New Yorkers and Californians of the macaque world. After a long period of exposure, the rhesus monkeys developed peacemaking skills on a par with those of their more tolerant counterparts. Even after separation from the stumptails, the rhesus showed nearly four times more friendly reunions following fights than is typical of their species. These new and improved rhesus monkeys confirmed the power of conformism.[39]

One of the most intriguing sides of social learning—defined as learning from others—is the secondary role of reward. While individual learning is driven by immediate incentives, such as a rat learning to press a lever to obtain food pellets, social learning doesn't work this way. Sometimes conformism even *reduces* rewards—after all, the vervet monkeys missed out on half of the available food. We once conducted an experiment in which capuchin monkeys watched a monkey model open one of three differently colored boxes. Sometimes the boxes contained food, but at other times they were empty. It didn't matter: the monkeys copied the model's choices regardless of whether there was any reward.[40]

There are even examples of social learning in which the benefits, instead of going to the performer, go to someone else. At the Mahale Mountains in Tanzania, I regularly saw a chimpanzee walk up to another, vigorously scratch the other's back with his or her fingernails, then settle down to groom the other. In between the grooming, more scratching might follow. This behavior has been known for a long time and has thus far been reported for only one other field site. It is a locally learned tradition, but here's the rub: when one scratches oneself, it is usually due to itching, and the act brings instant relief. In the case of the social scratch, however, the performer does not feel relief—the recipient does.[41]

Primates occasionally learn habits from others that do pay off, such as when chimpanzee youngsters learn to crack nuts with stones. But even then things are not as simple as they appear. Sitting next to their nut-cracking moms, infant chimps are total klutzes. They put nuts on top of stones, stones on top of nuts, and push them all together in a heap only to rearrange them over and over. They gain nothing from this playful activity. They also hit nuts with a hand, or stamp them hard with a foot, which fails to crack anything. Palm and panda nuts are far too tough for them. Only after three years of futile efforts do young chimps have enough coordination and strength to break open their first nut with a pair of stones, but they still have to wait until they are six or seven to reach adult skill levels.[42] Since they utterly fail at this task for so many years in a row, it is unlikely that food is the incentive. They may even experience negative consequences, such as smashed fingers. Yet young chimps happily persist, inspired by the example of their elders.

How little rewards matter is also evident from habits that lack benefits. In our own species, we have fads such as wearing a baseball cap backward or pants that hang low enough to impede locomotion. But in other primates, too, we find seemingly useless fashions and habits. A nice example is the N-family in a group of rhesus monkeys that I observed long ago at the Wisconsin Primate Center. This matriline was headed by an aging matriarch, Nose, all of whose offspring had names starting with the same letter, such as Nuts, Noodle, Napkin, Nina, and so on. Nose had developed the odd routine of drinking from a water basin by dipping her entire underarm into it, then licking her hand and the hair on her arm. Amusingly, all her offspring, and later her grand-children adopted the exact same technique. No other monkeys in the troop, or any other that I knew, drank like this, yet there was absolutely no advantage to it. It did not allow the N-family to access anything that other monkeys had no access to.

Or take the way chimpanzees sometimes develop local dialects, such as the excited grunts uttered while snacking on tasty food. These grunts differ not only from group to group but also per food type, such as a particular grunt heard only while they eat apples. When the Edinburgh Zoo introduced chimpanzees from a Dutch zoo to its residents, it took those others three years to get socially integrated. Initially, the newcom-ers uttered different grunts while eating apples, but by the end they con-verged on the same grunts as the locals. They had adjusted their calls so that they sounded more like those of the residents. While the media hyped this finding by saying that Dutch chimps had learned to speak Scottish, it was more like picking up an accent. The bonding between individuals of different backgrounds had resulted in conformism, even though chimps are not particularly known for vocal flexibility.[43]

Clearly, social learning is more about fitting in and acting like others than about rewards. This is why my book on animal culture was entitled *The Ape and the Sushi Master.* I chose this title partly to honor Imanishi and the Japanese scientists who gave us the animal culture concept, but also because of a story I had heard about how apprentice sushi masters learn their trade. The apprentice slaves in the shadow of the master of an art requiring rice of the right stickiness, precisely cut ingredients, and the eye-catching arrangements for which Japanese cuisine is famous. Anyone who has ever tried to cook rice, mix it with vinegar, and cool it off with a handheld fan so as to mold fresh rice balls in one's hands knows how complex a skill it is, and it is only a small part of the job. The apprentice learns mostly through passive observation. He washes the dishes, mops the floor, bows to the clients, fetches ingredients, and in the meantime follows from the corners of his eyes, without ever asking a question, everything the sushi master does. For three years he watches without being allowed to make actual sushi for the patrons of the restaurant: an extreme case of exposure without practice. He is waiting for the day when he will be invited to make his first sushi, which he will do with remarkable dexterity.

Whatever the truth about the sushi master's education, the point is that repeated observation of a skilled model firmly plants action sequences in one's head that come in handy, sometimes much later, when one needs to carry out the same task. Tetsuro Matsuzawa, who studied nut-cracking in West African chimpanzees, views social learning as based on a devoted master-apprentice relationship, in the same way that I developed my Bonding- and Identification-based Observational Learning model (BIOL).[44] Both views reject the traditional focus on incentives and replace it with one on social connections. Animals strive to act like others, especially others whom they trust and feel close to. Conformist biases shape society by promoting the absorption of habits and knowledge accumulated by previous generations. This by itself is obviously advantageous—and not just in the primates—so even though conformism is not driven by immediate benefits, it likely assists survival.

What's in a Name?

Konrad Lorenz was a big corvid fan. He always kept jackdaws, crows, and ravens around his house in Altenberg, near Vienna, and considered them the birds with the highest mental development. In the same way that I, as a student, took walks with my tame jackdaws flying above me, he traveled with Roah, his old raven and "close friend." And like my jackdaws, the raven would come down from the sky and try to make Lorenz follow by moving his tail sideways before him. It is a quick gesture that is not easily noticed from a distance yet hard to miss if done right in front of your face. Curiously, Roah used his own name to call Lorenz, whereas ravens normally call one another with a sonorous, deep-throated call-note described by Lorenz as a metallic "krackkrack-krack." Here is what he said about Roah's invitations:

> Roah bore down on me from behind, and, flying close over my head, he wobbled with his tail and then swept upwards again, at the same time looking backwards over his shoulder to see if I was following. In accompaniment of this sequence of movements Roah, instead of uttering the above described call-note, said his own name, with human intonation. The most peculiar thing about this was that Roah used the human word for me only. When addressing one of his own species, he employed the normal innate call-note.[45]

Lorenz denied that he had taught his raven to call like this—after all, he had never rewarded him for it. He suspected that Roah must have inferred that since "Roah!" was the call-note Lorenz used for him, it might also work in reverse. This sort of behavior may appear in animals that contact one another vocally and are moreover great imitators. As we shall see, this also holds for dolphins. In the primates, on the other hand, individual identity is usually visually determined. The face is the most characteristic part of the body; hence face recognition is highly developed and has been demonstrated in multiple ways in both monkeys and apes.

It is not just faces that they pay attention to, however. During our studies, we discovered how intimate chimps are with one another's der-rières. In one experiment, they first saw a picture of the behind of one of their group mates followed by two facial pictures. Only one of both faces belonged to the behind, however. Which one would they select on the touchscreen? It was a typical matching-to-sample task of the type invented by Nadia Kohts before the computer age. We found that our apes selected the correct portrait, the one that went with the butt they had seen. They were only successful, though, with chimps that they knew personally. That they failed with pictures of strangers suggests that it was not based on something in the pictures themselves, such as color or size. They must possess a whole-body image of familiar individuals, knowing them so well that they can connect any part of their body with any other part.

In the same way, we are able to locate friends and relatives in a crowd even if we only see their backs. Having published our findings under the suggestive title "Faces and Behinds," everyone thought it was funny that apes could do this, and we received an IgNobel prize for the study. This parody of the Nobel Prize honors research that "first makes people laugh, and then think."[46]

I do hope it makes people think, because individual recognition is the cornerstone of any complex society.[47] That animals have this capacity is often underestimated by humans, for whom all members of a given species look alike. Among themselves, however, animals generally have no trouble telling one another apart. Take dolphins, which for us are hard to identify because they all seem to have the same smiley face. Without equipment, we aren't privy to their main channel of communication, which is underwater sound. Investigators typically follow them around on the surface in a boat, as I did with my former student Ann Weaver, who recognizes about three hundred bottlenose dolphins in the Boca Ciega Bay Intracoastal Waterway estuary, in Florida. Ann carries an enormous photo album with close-ups of every dorsal fin in the area, which she has patrolled for over fifteen years. She visits the bay nearly every day in a small motorboat while on the lookout for surfacing dolphins. The dorsal fin is the body part we see most easily, and each one is shaped slightly differently. Some are tall and sturdy, while others hang to one side or miss a chunk due to fights or shark attacks.

From these identifications, Ann knew that some males form alliances and travel together all the time. They swim synchronously and surface together. The few times that they are not near each other, they get into trouble with rivals, who sense an opportunity. Females and calves, up to the age of five or six, move together, too. Otherwise dolphin society is *fission-fusion*, meaning that individuals gather in temporary combinations that vary from hour to hour and from day to day. Knowing who is around by looking at a small body part that regularly sticks out of the water is a rather cumbersome technique, however, compared to how dolphins themselves recognize one another.

Dolphins know one another's calls. This by itself is not so special, since we too recognize each other's voices, as do many other animals. The morphology of the vocal apparatus (mouth, tongue, vocal cords, lung capacity) varies greatly, which allows us to recognize voices by their pitch, loudness, and timbre. We have no trouble hearing the gender and age of a speaker or singer, but we also recognize individual voices. When I sit in my office and hear colleagues talking around the corner, I don't need to see them to know who they are.

Dolphins go much further, however. They produce *signature whistles*, which are high-pitched sounds with a modulation that is unique for each individual. Their structure varies the way ring-tone melodies vary. It is not so much the voice but the melody that marks them. Young dolphins develop personalized whistles in their first year. Females keep the same melody for the rest of their lives, whereas males adjust theirs to those of their closest buddies, so that the calls within a male alliance sound alike.[48] Dolphins utter signature whistles especially when they are isolated (lonely ones in captivity do so all the time) but also before aggregating in large groups in the ocean. At such moments, identities are broadcast frequently and widely, which makes sense in a fission-fusion species that dwells in murky water. That whistles are used for individual identification was shown by playing them back through underwater speakers. Dolphins pay more attention to sounds associated with close kin than to those of others. That this is based not on mere voice recognition but on the call's specific melody was demonstrated by playing back computer-generated sounds that mimicked the melodies: the voice was left out while the melody was preserved. These synthesized calls triggered the same responses as the originals.[49]

Dolphins have an incredible memory for their friends. The American animal behaviorist Jason Bruck took advantage of the fact that captive dolphins are regularly moved from one place to another for breeding purposes. He played back signature whistles of tank mates that had left long ago. In response to familiar calls, dolphins would become active, approach the speaker, and call in return. Bruck found that dolphins have no trouble recognizing former tank mates regardless of how much or little time they had spent together in the past or how long it had been since they had last seen them. The longest time interval in the study was when a female named Bailey recognized the whistles of Allie, a female she had lived with elsewhere twenty years before.[50]

Increasingly, experts view signature whistles as *names*. They are not just identifiers that individuals produce themselves but are sometimes mimicked. For dolphins, addressing specific companions by their own whistles is like calling them by name. While Roah used his own name to call Lorenz, dolphins sometimes mimic the characteristic call of someone else to draw his or her attention. That they do so is obviously hard to prove by observation alone; hence this issue was, again, addressed with playbacks. Working with bottlenose dolphins off the coast of Scotland, near the University of St. Andrews, Stephanie King and Vincent Janik recorded the signature whistles of free-ranging dolphins. They then played the calls back through a submerged speaker while the dolphins who had produced them still swam in the vicinity. The dolphins replied by calling back, sometimes multiple times, to their own characteristic whistles, as if confirming that they'd heard themselves being called.[51]

The deep irony of animals calling one another by name is, of course, that it was once taboo for scientists to name their animals. When Imanishi and his followers started doing so, they were ridiculed, as was Goodall when she gave her chimps names like David Greybeard and Flo. The complaint was that by using names we were humanizing our subjects. We were supposed to keep our distance and stay objective, and to never forget that only humans have names.

As it turns out, on this issue some animals may have been ahead of us.

9 | EVOLUTIONARY COGNITION

Given how easily we string the words *animal* and *cognition* together as if there were nothing to it—as if these words might even *belong* together!—it is hard to imagine the struggle we went through to reach this point. Some animals were considered good learners or hard-wired for clever solutions, but *cognition* was way too big a word for what they did. Even though for many people animal intelligence is self-evident, science never takes anything at face value. We want proof, which with regard to animal cognition has now become overwhelming—so much so, in fact, that we risk forgetting the immense resistance that we had to overcome. This is why I have paid ample attention to the history of our field. There were early pioneers, such as Köhler, Kohts, Tolman, and Yerkes, and a second generation, such as Menzel, Gallup, Beck, Shettleworth, Kummer, and Griffin. The third generation, to which I myself belong, includes so many evolutionary cognitivists that I am not going to list them here, but we too faced an uphill battle.

I can't count the number of times I have been called naïve, romantic, soft, unscientific, anthropomorphic, anecdotal, or just a sloppy thinker for proposing that primates follow political strategies, reconcile after fights, empathize with others, or understand the social world around them. Based on a lifetime of firsthand experience, none of these claims seemed particularly audacious to me. So one can imagine what happened to scientists suggesting awareness, linguistic capacities, or logical reasoning. Every claim was picked apart and held up against the light of alternative theories, which invariably sounded simpler given that they derived from the behavior of pigeons and rats in the confines of a Skinner box.

They were not always so simple, though—accounts based on associative learning can get quite convoluted compared to ones that merely postulate an extra mental faculty—but in those days, learning was thought to explain everything. Except, of course, when it didn't. In the latter case, we clearly hadn't thought long and hard enough about the issue at hand or we had failed to conduct the right experiments. At times, the wall of skepticism seemed more ideological than scientific, a bit the way we biologists feel about creationists. However compelling the data we bring to the table, they never suffice. Things must be believed to be seen, as Willy Wonka sang, and entrenched disbelief is oddly immune to evidence. The "slayers" of the cognitive view were not open to it.

This epithet comes from the American zoologist Marc Bekoff and the philosopher Colin Allen who early on picked up Griffin's torch for cognitive ethology. They divided attitudes toward animal cognition into three types: the slayers, the skeptics, and the proponents. When first writing about this in 1997, slayers were still abundant:

Slayers deny any possibility of success in cognitive ethology. In our analyses of their published statements, we have found that they sometimes conflate the difficulty of doing rigorous cognitive ethological investigations with the impossibility of doing so. Slayers also often ignore specific details of work by cognitive ethologists and frequently mount philosophically motivated objections to the possibility of learning anything about animal cognition. Slayers do not believe that cognitive ethological approaches can lead, and have led, to new and testable hypotheses. They often pick out the most difficult and least accessible phenomena to study (e.g. consciousness) and then conclude that because we can gain little detailed knowledge about this subject, we cannot do better in other areas. Slayers also appeal to parsimony in explanations of animal behavior, but they dismiss the possibility that cognitive explanations can be more parsimonious than noncognitive alternatives, and they deny the utility of cognitive hypotheses for directing empirical research.[1]

When Emil Menzel told me about the prominent professor—clearly a slayer—who tried to ambush him but ended up with his foot in his mouth, he added an interesting side note. The same professor publicly challenged young Menzel to tell him what capacities he could possibly hope to find in apes that were not also present in pigeons. In other words, why waste your time on those willful, hard-to-control apes if animal intelligence is essentially the same across the board?

While this was the prevailing attitude at the time, the field has come around to a much more evolutionary approach, which recognizes that every species has a different cognitive story to tell. Each organism has its own ecology and lifestyle, its own *Umwelt*, which dictates what it needs to know in order to make a living. There is not a single species that can stand model for all the others, most certainly not one with a brain as tiny as a pigeon's. Pigeons are plenty intelligent, but size does matter. Brains are the most "expensive" organs around. They are true energy hogs, using twenty times more calories per unit than muscle tissue. Menzel could simply have countered that since ape brains are several hundred times heavier than those of pigeons and hence burn vastly more energy, it stands to reason that apes face greater cognitive challenges. Otherwise mother nature indulged in a shocking extravagance, something she is not known for. In the utilitarian view of biology, animals have the brains they need—nothing more, nothing less. Even *within* a species, the brain may change depending on how it is being used, such as the way song-related areas seasonally expand and contract in the songbird brain.[2] Brains adapt to ecological requirements, as does cognition.

We have also met a second type of slayer, though, and they have been even harder to deal with since they don't share an interest in animal behavior. All they care about is humanity's position in the cosmos, which science has been undercutting since the days of Copernicus. Their struggle has become rather hopeless, though, because if there is one overall trend in our field, it is that the wall between human and animal cognition has begun to resemble a Swiss Gruyère full of holes. Time after time we have demonstrated capacities in animals that were thought to set our species apart. Proponents of human uniqueness face the possibility that they have either grossly overestimated the complexity of what humans do or underestimated the capacities of other species.

Neither possibility is a pleasant thought, because their deeper problem is evolutionary continuity. They can't stand the notion of humans as modified apes. Like Alfred Russel Wallace, they feel that evolution must have skipped the human head. Although this view is currently on its way out in psychology, which under the sway of neuroscience is edging ever closer to the natural sciences, it is still prevalent in the humanities and most of the social sciences. Typical is a recent reaction by the American anthropologist Jonathan Marks to the overwhelming evidence that animals pick up habits from one another, hence show cultural variability: "Labeling ape behavior as 'culture' simply means you have to find another word for what humans do."[3]

How much more refreshing was David Hume, the Scottish philosopher who held animals in such high esteem that he wrote that "no truth appears to me more evident than that beasts are endow'd with thought and reason as well as men." In line with my position throughout this book, Hume summarized his view in the following principle:

> *'Tis from the resemblance of the external actions of animals to those we ourselves perform, that we judge their internal likewise to resemble ours; and the same principle of reasoning, carry'd one step farther, will make us conclude that since our internal actions resemble each other, the causes, from*

which they are deriv'd, must also be resembling. When any hypothesis, therefore, is advanc'd to explain a mental operation, which is common to men and beasts, we must apply the same hypothesis to both.[4]

Formulated in 1739, more than a century before Darwin's theory saw the light, Hume's Touchstone offers a perfect starting point for evolutionary cognition. The most parsimonious assumption we can make about behavioral and cognitive similarities between related species is that they reflect shared mental processes. Continuity ought to be the default position for at least all mammals, and perhaps also birds and other vertebrates.

When this view finally gained the upper hand about twenty years ago, supportive evidence poured in from all sides. It was not just the primates anymore but also the canines, corvids, elephants, dolphins, parrots, and so on. The stream of discoveries became unstoppable, featured in the media on a weekly basis to the point that *The Onion* felt like spoofing the trend in an article claiming that dolphins are not nearly as smart on land as they are in the ocean.[5] Joking aside, this was a valid point related to the species-appropriate testing that is one of our field's main challenges. The public got used to a great variety of claims, including news stories and blogs about animals liberally sprinkled with terms like *thinking, sentience,* and *rational.*

Some of it was hype, but many reports presented serious peer-reviewed studies based on years of painstaking research. As a result, evolutionary cognition began to gain standing and attract a growing influx of students ready to cut their teeth on a promising topic. Students like nothing better than a new area where fresh ideas matter. Nowadays many scientists studying animal behavior proudly put the word *cognitive* in statements about their research, and scientific journals add this trendy term to their names, realizing that it attracts more readers than any other in behavioral biology. The cognitive view has won.

But an assumption is still only an assumption. It doesn't absolve us from working hard on the issues at hand, which is to determine at what cognitive level a given species operates and how this suits its ecology and lifestyle. What are its cognitive strengths, and how do these relate to survival? It all goes back to the kittiwake story: some species need to recognize their young and others just don't. The first will pay attention to individual identities, while the second can safely ignore them. Or recall how Garcia's nauseated rats broke the rules of operant conditioning, as if to drive home the point that remembering toxic food is a magnitude more important than knowing which bar delivers pellets. Animals learn what they need to learn and have specialized ways of sifting through the massive information around them. They actively seek, collect, and store information. They are often incredibly good at one particular task, such as caching and remembering food items or fooling predators, whereas some species are endowed with the brainpower to tackle a wide array of problems.

Cognition may even push physical evolution in a particular direction, such as the reliance of New Caledonian crows on tools crafted out of leaves and twigs. These crows have straighter bills than other corvids and also more forward-facing eyes. The bill shape helps them get a stable grip on their tools, whereas binocular vision lets them peer deeply into the crevices from which they extract caterpillars.[6] Cognition is not merely a product of an animal's senses, anatomy, and brainpower, therefore, but the relation also works the other way around. Physical features adapt to an animal's cognitive specializations. The human hand may be another example, having evolved its fully opposable thumbs and remarkable versatility to suit our reliance on refined tools, from stone axes to the modern smartphones. This is why evolutionary cognition is such a perfect label for our field, because only evolutionary theory can make sense of survival, ecology, anatomy, and cognition all at once. Instead of searching for a general theory that covers all cognition on the planet, it treats every species as a case study. Of course, some cognitive principles are common to all organisms, but we don't seek to downplay variation between species with lifestyles, ecologies, and *Umwelten* as different as, say, a dolphin and a dingo or a macaw and a monkey. Each one faces its own specific cognitive challenges.

Once comparative psychologists began to appreciate that every species is special, and that learning is dictated by biology, they gradually began to enter the fold of evolutionary cognition. Their discipline greatly contributed to it through its long history of carefully controlled experiments and its many scientists with cognitive leanings. Even though these pioneers worked mostly under the radar and were forced to publish in second-tier journals, they described "higher mental processes" that they felt excluded learning.[7] Given the absolute hegemony of behaviorism at the time, it made sense to define cognition in opposition to learning, but this always strikes me as a mistake. This dichotomy is as false as the one that pits nature against nurture. The reason we rarely talk about instincts anymore is that nothing is purely genetic: the environment always plays a role. In the same way, pure cognition is a figment of the imagination. Where would cognition be without learning? Some sort of information gathering is always part of it. Even Köhler's apes, which heralded the study of animal cognition, had previous experience with boxes and sticks. Rather than looking at the cognitive revolution as a blow to learning theory, therefore, it is more like a marriage. The relationship has had its ups and downs, but in the end, learning theory will survive within the framework of evolutionary cognition. In fact, it will be an essential part of it.

The same holds for ethology. Its ideas about behavioral evolution are far from dead. They live on in many areas of science together with the ethological method. Systematic description and observation of behavior are at the core of all animal fieldwork as well as studies of child behavior, mother-infant interactions, nonverbal communication, and so on. The study of human emotions treats facial expressions as fixed action patterns while relying on the ethological method to measure them. For this reason, I don't look at the current flowering of evolutionary cognition as a break with the past but rather as a moment in time when forces and approaches that have been around for a century or longer have won the upper hand. We finally have the breathing room to discuss the marvelous ways in which animals gather and organize information. And while the slayers of the cognitive view are a dying breed, we obviously still have the other two categories around—the skeptics and proponents—both of whom are essential. As a proponent myself, I do appreciate my more skeptical colleagues. They keep us on our toes and force us to design clever experiments to answer their questions. So long as progress is our shared goal, this is exactly how science ought to work.

Even though the study of animal cognition is often portrayed as an attempt to find out "what they think," that is not really what it is all about. We're not after private states and experiences, although it would be great if one day we could know more about them. For the moment, our goal is more modest: we wish to pinpoint proposed mental processes by measuring observable outcomes. In this sense, our field is no different from other scientific endeavors, from evolutionary biology to physics. Science always starts with a hypothesis, followed by the testing of its predictions. If animals plan ahead, they should retain tools that they will need later on. If they understand cause-effect relations, they should avoid the trap in the trap-tube the first time they encounter it. If they know what others know, they should vary their behavior depending on what they have seen others pay attention to. If they have political talents, they should treat the friends of their rivals with circumspection. Having discussed dozens of such predictions, and the experiments and observations they have inspired, the pattern of research is obvious. Generally, the more lines of evidence converge in support of a given mental faculty, the stronger it stands. If planning for the future is evident in everyday behavior, in tests with delayed tool use, as well as in untrained food caching and foraging choices, we are in pretty good shape to claim that at least some species have this capacity.

But still I often feel that we are too obsessed with the pinnacles of cognition, such as theory of mind, self-awareness, language, and so on, as if making grandiose claims about these is all that matters. It is time for our field to move away from interspecific bragging contests (my crows are smarter than your monkeys) and the black-and-white thinking it engenders. What if theory of mind rests not on one big capacity but on an entire set of smaller ones? What if self-awareness comes in gradations? Skeptics often urge us to break down larger mental concepts by asking what exactly we mean. If we mean less than we claim, they wonder why we don't use a more reduced, down-to-earth description of the phenomenon.

I have to agree. We should start focusing on the processes behind higher capacities. They often rest on a wide range of cognitive mechanisms, some of which may be shared by many species, while others may be fairly restricted. We went through all this in the discussion of social reciprocity, which was initially conceived as animals remembering specific favors in order to repay them. Many scientists were unwilling to assume that monkeys, let alone rats, kept tabs on every social interaction. We now realize that this is not a requirement for tit-for-tat, and that not only animals, but also humans often exchange favors on a more basic, automated level related to long-term social ties. We help our buddies, and our buddies help us, but we aren't necessarily counting.[8] Ironically, the study of animal cognition not only raises the esteem in which we hold other species, but also teaches us not to overestimate our own mental complexity.

We urgently need a bottom-up view that focuses on the building blocks of cognition.[9] This approach will also need to include the emotions—a topic I have barely touched upon but that is close to my heart and is in equal need of attention. Breaking down mental capacities into all of these components may lead to less spectacular headlines, but our theories will be more realistic and informative as a result. It will also require a greater involvement of neuroscience. At the moment, its role is rather limited. Neuroscience may tell us where things happen in the brain, but this hardly helps us formulate new theories or design insight-ful tests. But while the most interesting work in evolutionary cognition is still mostly behavioral, I am sure this is going to change. Neuro-science has thus far only scratched the surface. In the coming decades, it will inevitably become less descriptive and more theoretically relevant to our discipline. In time, a book such as the present one will have a huge amount of neuroscience in it, explaining which brain mechanisms are responsible for the behavior observed.

This will be an excellent way to test the continuity assumption, since homologous cognitive processes imply shared neural mechanisms. Such evidence is already accumulating for face recognition in monkeys and humans, the processing of rewards, the role of the hippocampus in memory and of mirror neurons in imitation. The more evidence for shared neural mechanisms we find, the stronger the argument for homology and continuity will become. And conversely, if two species engage different neural circuits to achieve similar outcomes, the conti-nuity stance will need to be abandoned in favor of one based on conver-gent evolution. The latter is quite powerful, too, having produced face recognition in both primates and wasps, for example, or flexible tool use in both primates and corvids.

The study of animal behavior is among the oldest of human endeav-ors. As hunter-gatherers, our ancestors needed intimate knowledge of flora and fauna, including the habits of their prey. Hunters exercise min-imal control: they anticipate the moves of animals and are impressed by their cunning if they escape. They also need to watch their back for species that prey on them. The human-animal relationship was rather egalitarian during this time. A more practical knowledge became nec-essary when our ancestors took up agriculture and began to domesti-cate animals for food and muscle power. Animals became dependent on us and subservient to our will. Instead of anticipating their moves, we began to dictate them, while our holy books spoke of our domin-ion over nature. Both of these radically different attitudes—the hunter's and the farmer's—are recognizable in the study of animal cognition today. Sometimes we watch what animals do of their own accord, while at other times we put them in situations where they can do little else besides what we want them to do.

With the rise of a less anthropocentric orientation, however, the second approach may be on the decline, or at least add significant degrees of freedom. Animals should be given a chance to express their natural behavior. We are developing a greater interest in their variable lifestyles. Our challenge is to think more like them, so that we open our minds to their specific circumstances and goals and observe and understand them on their own terms. We are returning to our hunting ways, albeit more in the way that a wildlife photographer relies on the hunting instinct: not to kill but to reveal. Nowadays experiments often revolve around natural behavior, from courtship and foraging to prosocial attitudes. We seek ecological validity in our studies and follow the advice of Uexküll, Lorenz, and Imanishi, who encouraged human empathy as a way to understand other species. True empathy is not self-focused but other-oriented. Instead of making humanity the measure of all things, we need to evaluate other species by what *they* are. In doing so, I am sure we will discover many magic wells, including some as yet beyond our imagination.

NOTES

PROLOGUE

1 Charles Darwin (1972 [orig. 1871]), p. 105.
2 Ernst Mayr (1982), p. 97.
3 Richard Byrne (1995), Jacques Vauclair (1996), Michael Tomasello and Josep Call (1997), James Gould and Carol Grant Gould (1999), Marc Bekoff et al. (2002), Susan Hurley and Matthew Nudds (2006), John Pearce (2008), Sara Shettleworth (2012), and Clive Wynne and Monique Udell (2013).

CHAPTER 1: MAGIC WELLS

1 Werner Heisenberg (1958), p. 26.
2 Jakob von Uexküll (1957 [orig. 1934]), p. 76. See also Jakob von Uexküll (1909).
3 Thomas Nagel (1974).
4 Ludwig Wittgenstein (1958 [orig. 1953]), p. 225.
5 Martin Lindauer (1987), p 6, quoting Karl von Frisch.
6 Donald Griffin (2001).
7 Ronald Lanner (1996).
8 Niko Tinbergen, (1953), Eugène Marais (1969), Dorothy Cheney and Robert Seyfarth (1992), Alexandra Horowitz (2010), and E. O. Wilson (2010).
9 Benjamin Beck (1967).
10 Preston Foerder et al. (2011).
11 Daniel Povinelli (1989).

12 Joshua Plotnik et al. (2006).

13 Lisa Parr and Frans de Waal (1999).

14 Doris Tsao et al. (2008).

15 Konrad Lorenz (1981), p. 38.

16 Edward Thorndike (1898) inspired Edwin Guthrie and George Horton (1946).

17 Bruce Moore and Susan Stuttard (1979).

18 Edward Wasserman (1993).

19 Donald Griffin (1976).

20 Victor Stenger (1999).

21 Jan van Hooff (1972), Marina Davila Ross et al. (2009).

22 Frans de Waal (1999).

23 Gordon Burghardt (1991).

24 Frans de Waal (2000), Nicola Koyama (2001), Mathias Osvath and Helena Osvath (2008).

25 William Hodos and C. B. G. Campbell (1969).

26 "Pigeon, rat, monkey, which is which? It doesn't matter." B. F. Skinner (1956), p. 230.

27 Konrad Lorenz (1941).

CHAPTER 2: A TALE OF TWO SCHOOLS

1 Esther Cullen (1957).

2 Bonnie Perdue et al. (2011), Steven Gaulin and Randall Fitzgerald (1989).

3 Bruce Moore (1973), Michael Domjan and Bennett Galef (1983).

4 Sara Shettleworth (1993), Bruce Moore (2004).

5 Louise Buckley et al. (2011).

6 Harry Harlow (1953), p. 31.

7 Donald Dewsbury (2006), p. 226.

8 John Falk (1958).

9 Keller Breland and Marian Breland (1961).

10 B. F. Skinner (1969), p. 40.

11 William Thorpe (1979).

12 Richard Burkhardt (2005).

13 Desmond Morris (2010), p. 51.

14 Anne Burrows et al. (2006).

15 George Romanes (1882), George Romanes (1884).

16 C. Lloyd Morgan (1894), pp. 53–54.
17 Roger Thomas (1998), Elliott Sober (1998).
18 C. Lloyd Morgan (1903).
19 Frans de Waal (1999).
20 René Röell (1996).
21 Niko Tinbergen (1963).
22 Oskar Pfungst (1911).
23 Douglas Candland (1993).
24 "The Remarkable Orlov Trotter," Black River Orlovs, www.infohorse .com/ShowAd.asp?id=3693.
25 Juliane Kaminski et al. (2004).
26 Gordon Gallup (1970).
27 Robert Epstein et al. (1981).
28 Roger Thompson and Cynthia Contie (1994), but see Emiko Uchino and Shigeru Watanabe (2014).
29 Celia Heyes (1995).
30 Daniel Povinelli et al. (1997).
31 Jeremy Kagan (2000), Frans de Waal (2009a).
32 Kinji Imanishi (1952), Junichiro Itani and Akisato Nishimura (1973).
33 Bennett Galef (1990).
34 Frans de Waal (2001).
35 Satoshi Hirata et al. (2001).
36 David Premack and Ann Premack (1994).
37 Josep Call (2004), Juliane Bräuer et al. (2006)
38 Josep Call (2006).
39 Daniel Lehrman (1953).
40 Richard Burkhardt (2005), p. 390.
41 Ibid., p. 370; Hans Kruuk (2003).
42 Frank Beach (1950).
43 Donald Dewsbury (2000).
44 John Garcia et al. (1955).
45 Shettleworth (2010).
46 Hans Kummer et al. (1990).
47 Frans de Waal (2003b).
48 Hans Kruuk (2003), p. 157.
49 Niko Tinbergen and Walter Kruyt (1938).
50 Frans de Waal (2007 [orig. 1982]).

CHAPTER 3: COGNITIVE RIPPLES

1 Wolfgang Köhler (1925). The German original, *Intelligenzprüfungen an Anthropoiden*, appeared in 1917.
2 Robert Yerkes (1925), p. 120.
3 Robert Epstein (1987).
4 Emil Menzel (1972). Menzel was interviewed by the author in 2001.
5 Jane Goodall (1986), p. 357.
6 Frans de Waal (2007 [orig. 1982]).
7 Jennifer Pokorny and Frans de Waal (2009).
8 John Marzluff and Tony Angell (2005), p. 24.
9 John Marzluff et al. (2010); Garry Hamilton (2012).
10 Michael Sheehan and Elizabeth Tibbetts (2011).
11 Johan Bolhuis and Clive Wynne (2009), see also Frans de Waal (2009a).
12 Marco Vasconcelos et al. (2012).
13 Jonathan Buckley et al. (2010).
14 Barry Allen (1997).
15 M. M. Günther and Christophe Boesch (1993).
16 Gen Yamakoshi (1998).
17 "Tool use is the external employment of an unattached environmental object to alter more efficiently the form, position, or condition of another object, another organism, or the user itself when the user holds or carries the tool during or just prior to use and is responsible for the proper and effective orientation of the tool." Benjamin Beck (1980), p. 10.
18 Robert Amant and Thomas Horton (2008).
19 Jane Goodall (1967), p. 32.
20 Crickette Sanz et al. (2010).
21 Christophe Boesch et al. (2009), Ebang Wilfried and Juichi Yamagiwa (2014).
22 William McGrew (2010).
23 Jill Pruetz and Paco Bertolani (2007).
24 Tetsuro Matsuzawa (1994), Noriko Inoue-Nakamura and Tetsuro Matsuzawa (1997).
25 Jürgen Lethmate (1982).
26 Carel van Schaik et al. (1999).
27 Thibaud Gruber et al. (2010), Esther Herrmann et al. (2008).
28 Thomas Breuer et al. (2005), Jean-Felix Kinani and Dawn Zimmerman (2015).

29 Eduardo Ottoni and Massimo Mannu (2001).

30 Dorothy Fragaszy et al. (2004).

31 Julio Mercader et al. (2007).

32 Elisabetta Visalberghi and Luca Limongelli (1994).

33 Luca Limongelli et al. (1995), Gema Martin-Ordas et al. (2008).

34 William Mason (1976), pp. 292–93.

35 Michael Gumert et al. (2009).

36 "Honey Badgers: Masters of Mayhem," *Nature*, broadcast Feb. 19, 2014, Public Broadcasting Service.

37 Alex Weir et al. (2002).

38 Gavin Hunt (1996), Hunt and Russell Gray (2004).

39 Christopher Bird and Nathan Emery (2009), Alex Taylor and Russell Gray (2009), Sarah Jelbert et al. (2014).

40 Alex Taylor et al. (2014).

41 Natacha Mendes et al. (2007), Daniel Hanus et al. (2011).

42 Daniel Hanus et al. (2011).

43 Gavin Hunt et al. (2007), p. 291.

44 William McGrew (2013).

45 Alex Taylor et al. (2007).

46 Nathan Emery and Nicola Clayton (2004).

47 Vladimir Dinets et al. (2013).

48 Julian Finn et al. (2009).

CHAPTER 4: TALK TO ME

1 Bishop of Polignac, cited in Corbey (2005), p. 54.

2 Nadezhda Ladygina-Kohts (2002 [orig. 1935]).

3 Herbert Terrace et al. (1979).

4 Irene Pepperberg (2008).

5 Michele Alexander and Terri Fisher (2003).

6 Norman Malcolm (1973), p. 17.

7 Jerry Fodor (1975), p. 56.

8 Irene Pepperberg (1999).

9 Bruce Moore (1992)

10 Alice Auersperg et al. (2012).

11 Ewen Callaway (2012).

12 Sarah Boysen and Gary Berntson (1989).

13 Irene Pepperberg (2012).

14 Irene Pepperberg (1999), p. 327.

15 Sapolsky (2010).

16 Evolution of Language International Conferences, www.evolang.org.

17 Frans de Waal (2007 [orig. 1982], de Waal (1996), de Waal (2009a).

18 Dorothy Cheney and Robert Seyfarth (1990).

19 Kate Arnold and Klaus Zuberbühler (2008).

20 Toshitaka Suzuki (2014).

21 Brandon Wheeler and Julia Fischer (2012).

22 Tabitha Price (2013), Nicholas Ducheminsky et al. (2014).

23 Amy Pollick and Frans de Waal (2007), Katja Liebal et al. (2013), Catherine Hobaiter and Richard Byrne (2014).

24 Frans de Waal (2003a).

25 In 1980 Thomas Sebeok and the New York Academy of Sciences organized a conference entitled "The Clever Hans Phenomenon: Communication with Horses, Whales, Apes, and People."

26 Sue Savage-Rumbaugh and Roger Lewin (1994), p. 50, Jean Aitchison (2000).

27 Muhammad Spocter et al. (2010).

28 Sandra Wohlgemuth et al. (2014).

29 Andreas Pfenning et al. (2014).

30 Frans de Waal (1997), p. 38.

31 Robert Yerkes (1925), p. 79.

32 Oliver Sacks (1985).

33 Robert Yerkes (1943).

34 Vilmos Csányi (2000), Alexandra Horowitz (2009), Brian Hare and Vanessa Woods (2013).

35 Tiffani Howell et al. (2013).

36 Sally Satel and Scott Lilienfeld (2013).

37 Craig Ferris et al. (2001), John Marzluff et al. (2012).

38 Gregory Berns (2013).

39 Gregory Berns et al. (2013).

CHAPTER 5: THE MEASURE OF ALL THINGS

1 Sana Inoue and Tetsuro Matsuzawa (2007), Alan Silberberg and David Kearns (2009), Tetsuro Matsuzawa (2009).

2 Jo Thompson (2002).

3 David Premack (2010), p. 30.

4 Marc Hauser interviewed by Jerry Adler (2008).

5 The Public Broadcasting Service entitled a 2010 series *The Human Spark*.

6 Alfred Russel Wallace (1869), p. 392.

7 Suzana Herculano-Houzel et al. (2014), Ferris Jabr (2014).

8 Katerina Semendeferi et al. (2002), Suzana Herculano-Houzel (2009), Frederico Azevedo et al. (2009).

9 Ajit Varki and Danny Brower (2013), Thomas Suddendorf (2013), Michael Tomasello (2014).

10 Jeremy Taylor (2009), Helene Guldberg (2010).

11 Virginia Morell (2013), p. 232.

12 Robert Sorge et al. (2014).

13 Emil Menzel (1974).

14 Katie Hall et al. (2014).

15 David Premack and Guy Woodruff (1978).

16 Frans de Waal (2008), Stephanie Preston (2013).

17 Adam Smith (1976 [orig. 1759]), p. 10.

18 J. B. Siebenaler and David Caldwell (1956), p. 126.

19 Frans de Waal (2005), p. 191.

20 Frans de Waal (2009a).

21 Shinya Yamamoto et al. (2009).

22 Yuko Hattori et al. (2012).

23 Henry Wellman et al. (2000).

24 Ljerka Ostojić et al. (2013).

25 Daniel Povinelli (1998).

26 Derek Penn and Daniel Povinelli (2007).

27 David Leavens et al. (1996), Autumn Hostetter et al. (2001).

28 Catherine Crockford et al. (2012), Anne Marijke Schel et al. (2013).

29 Brian Hare et al. (2001).

30 Hika Kuroshima et al. (2003), Anne Marije Overduin-de Vries et al. (2013).

31 Anna Ilona Roberts et al. (2013).

32 Daniel Povinelli (2000).

33 Esther Herrmann et al. (2007).

34 Yuko Hattori et al. (2010).

35 Allan Gardner et al. (2011).

36 Frans de Waal (2001), de Waal et al. (2008), Christophe Boesch (2007).

37 Nathan Emery and Nicky Clayton (2001).

38 Thomas Bugnyar and Bernd Heinrich (2005); see also "Quoth the Raven," *Economist*, May 13, 2004.

39 Josep Call and Michael Tomasello (2008).

40 Atsuko Saito and Kazutaka Shinozuka (2013), p. 689.

41 Brian Hare et al. (2002), Ádám Miklósi et al. (2003), Hare and Michael Tomasello (2005), Monique Udell et al. (2008, 2010), Márta Gácsi et al. (2009).

42 Miho Nagasawa et al. (2015).

43 Leslie White (1959), p. 5.

44 Edward Thorndike (1898), p. 50, Michael Tomasello and Josep Call (1997).

45 Michael Tomasello et al. (1993ab), David Bjorklund et al. (2000).

46 Victoria Horner and Andrew Whiten (2005).

47 David Premack (2010).

48 Andrew Whiten et al. (2005), Victoria Horner et al. (2006), Kristin Bonnie et al. (2006), Horner and Frans de Waal (2010), Horner and de Waal (2009).

49 Michael Huffman (1996), p. 276.

50 Edwin van Leeuwen et al. (2014).

51 William McGrew and Caroline Tutin (1978).

52 Frans de Waal (2001), de Waal and Kristin Bonnie (2009).

53 Elizabeth Lonsdorf et al. (2004)

54 Victoria Horner et al. (2010), Rachel Kendal et al. (2015).

55 Christine Caldwell and Andrew Whiten (2002).

56 Friederike Range and Zsófia Virányi (2014).

57 Jeremy Kagan (2004), David Premack (2007).

58 Charles Darwin, Notebook M,1838, http://darwin-online.org.uk.

59 Lydia Hopper et al. (2008).

60 Frans de Waal (2009a), Delia Fuhrmann et al. (2014).

61 Suzana Herculano-Houzel et al. (2011, 2014).

62 Josef Parvizi (2009).

63 Robert Barton (2012).

64 Michael Corballis (2002), William Calvin (1982).

65 Natasja de Groot et al. (2010).

66 The "Mens vs aap—experiment" video can be seen at http://bit.ly/1gb LiCm.

67 Christopher Martin et al. (2014).

68 Frans de Waal (2007 [orig. 1982]).

69 Benjamin Beck (1982).

70 Alaska governor Sarah Palin, policy speech, Pittsburgh, PA, October 24, 2008.

CHAPTER 6: SOCIAL SKILLS

1 Frans de Waal (2007 [orig. 1982]).
2 Donald Griffin (1976).
3 Hans Kummer (1971), Kummer (1995).
4 Jane Goodall (1971).
5 Christopher Martin et al. (2014).
6 Frans de Waal and Jan van Hooff (1981).
7 Frans de Waal (2007 [orig. 1982]).
8 Marcel Foster et al. (2009)
9 Toshisada Nishida et al. (1992).
10 Toshisada Nishida (1983), Nishida and Kazuhiko Hosaka (1996).
11 Victoria Horner et al. (2011).
12 Malini Suchak and Frans de Waal (2012).
13 Hans Kummer et al. (1990), Frans de Waal (1991).
14 Richard Byrne and Andrew Whiten (1988).
15 Robin Dunbar (1998b).
16 Thomas Geissmann and Mathias Orgeldinger (2000).
17 Sarah Gouzoules et al. (1984).
18 Dorothy Cheney and Robert Seyfarth (1992).
19 Susan Perry et al. (2004).
20 Susan Perry (2008), p. 47.
21 Katie Slocombe and Klaus Zuberbühler (2007).
22 Dorothy Cheney and Robert Seyfarth (1986, 1989), Filippo Aureli et al. (1992).
23 Peter Judge (1991), Judge and Sonia Mullen (2005).
24 Ronald Schusterman et al. (2003).
25 Dalila Bovet and David Washburn (2003), Regina Paxton et al. (2010).
26 Jorg Massen et al. (2014a).
27 Meredith Crawford (1937).
28 Kim Mendres and Frans de Waal (2000).
29 Alicia Melis et al. (2006a), Alicia Melis et al. (2006b), Sarah Brosnan et al. (2006).
30 Frans de Waal and Michelle Berger (2000).
31 Ernst Fehr and Urs Fischbacher (2003).
32 Robert Boyd (2006), countered by Kevin Langergraber et al. (2007).
33 Malini Suchak and Frans de Waal (2012), Jingzhi Tan and Brian Hare (2013).

34 National Academies of Sciences and Engineering, Keck Futures Initiative Conference, Irvine, CA, November 2014.

35 E. O. Wilson (1975).

36 Michael Tomasello (2008), Gary Stix (2014), p. 77.

37 Emil Menzel (1972).

38 Joshua Plotnik et al. (2011).

39 Ingrid Visser et al. (2008).

40 Christophe Boesch and Hedwige Boesch-Achermann (2000).

41 The two photographs are featured in Gary Stix (2014).

42 Malini Suchak et al. (2014).

43 Michael Wilson et al. (2014).

44 Sarah Calcutt et al. (2014).

45 Hal Whitehead and Luke Rendell (2015).

46 Sarah Brosnan and Frans de Waal (2003). See also "Two Monkeys Were Paid Unequally," TED Blog Video, http://bit.ly/1GO05tz.

47 Sarah Brosnan et al. (2010), Proctor et al. (2013).

48 Frederieke Range et al. (2008), Claudia Wascher and Thomas Bugnyar (2013), Sarah Brosnan and Frans de Waal (2014).

49 Redouan Bshary and Ronald Noë (2003).

50 Redouan Bshary et al. (2006).

51 Alexander Vail et al. (2014).

52 Toshisada Nishida and Kazuhiko Hosaka (1996).

53 Jorg Massen et al. (2014b).

54 Caitlin O'Connell (2015).

CHAPTER 7: TIME WILL TELL

1 Robert Browning (2006 [orig. 1896]), p. 113.

2 Otto Tinklepaugh (1928).

3 Gema Martin-Ordas et al. (2013).

4 Marcel Proust (1913), p. 48.

5 Karline Janmaat et al. (2014), Simone Ban et al. (2014).

6 Endel Tulving (1972, 2001).

7 Nicola Clayton and Anthony Dickinson (1998).

8 Stephanie Babb and Jonathon Crystal (2006).

9 Sadie Dingfelder (2007), p. 26.

10 Thomas Suddendorf (2013), p. 103.

11 Endel Tulving (2005).

12 Mathias Osvath (2009).

13 Lucia Jacobs and Emily Liman (1991).

14 Nicholas Mulcahy and Josep Call (2006).

15 Mathias Osvath and Helena Osvath (2008), Osvath and Gema Martin-Ordas (2014).

16 Juliane Bräuer and Josep Call (2015).

17 Caroline Raby et al. (2007), Sérgio Correia et al. (2007), William Roberts (2012).

18 Nicola Koyama et al. (2006).

19 Carel van Schaik et al. (2013).

20 Anoopum Gupta et al. (2010), Andrew Wikenheiser and David Redish (2012).

21 Sara Shettleworth (2007), Michael Corballis (2013).

22 In 2011 French media compared Dominique Strauss-Kahn to a *"chimpanzé en rut."*

23 Richard Byrne (1995), p. 133, Robin Dunbar (1998a).

24 Ramona Morris and Desmond Morris (1966).

25 Philip Kitcher (2006), p. 136.

26 Harry Frankfurt (1971), p. 11, also Roy Baumeister (2008).

27 Jessica Bramlett et al. (2012).

28 Michael Beran (2002), Theodore Evans and Beran (2007).

29 Friederike Hilleman et al. (2014)

30 Adrienne Koepke et al. (in press).

31 Walter Mischel and Ebbe Ebbesen (1970).

32 David Leavens et al. (2001).

33 Walter Mischel et al. (1972), p. 217.

34 Michael Beran (2015).

35 Sarah Boysen and Gary Berntson (1995).

36 Edward Tolman (1927).

37 David Smith et al. (1995).

38 Robert Hampton (2004).

39 Allison Foote and Jonathon Crystal (2007).

40 Arii Watanabe et al. (2014)

41 Josep Call and Malinda Carpenter (2001), Robert Hampton et al. (2004).

42 Alastair Inman and Sara Shettleworth (1999).

43 *The Cambridge Declaration on Consciousness*, July 7, 2012, Francis Crick Memorial Conference at Churchill College, University of Cambridge.

CHAPTER 8: OF MIRRORS AND JARS

1 Joshua Plotnik et al. (2006). See also "Mirror Self-Recognition in Asian Elephants" (video), Jan. 11, 2015, http://bit.ly/1spFNoA.
2 Joshua Plotnik et al. (2014).
3 Michael Garstang et al. (2014).
4 Ulric Neisser (1967), p. 3.
5 Lucy Bates et al. (2007).
6 Karen McComb et al. (2014).
7 Karen McComb et al. (2011).
8 Joseph Soltis et al. (2014).
9 Gordon Gallup Jr. (1970), James Anderson and Gallup (2011).
10 Daniel Povinelli (1987).
11 Emanuela Cenami Spada et al. (1995), Mark Bekoff and Paul Sherman (2003).
12 Matthew Jorgensen et al. (1995), Koji Toda and Shigeru Watanabe (2008).
13 Doris Bischof-Köhler (1991), Carolyn Zahn-Waxler et al. (1992), Frans de Waal (2008).
14 Abigail Rajala et al. (2010), Liangtang Chang et al. (2015)
15 Frans de Waal et al. (2005).
16 Philippe Rochat (2003).
17 Diana Reiss and Lori Marino (2001).
18 Helmut Prior et al. (2008).
19 My translation of Jürgen Lethmate and Gerti Dücker (1973), p. 254.
20 Ralph Buchsbaum et al. (1987 [orig. 1938]).
21 Roland Anderson and Jennifer Mather (2010).
22 Katherine Harmon Courage (2013), p. 115.
23 Roland Anderson et al. (2010).
24 Jennifer Mather et al. (2010), Roger Hanlon and John Messenger (1996).
25 Roland Anderson et al. (2002).
26 Aristotle (1991), p. 323.
27 Jennifer Mather and Roland Anderson (1999), Sarah Zylinski (2015).
28 Roger Hanlon (2007), Hanlon (2013).
29 Roger Hanlon et al. (1999).
30 Culum Brown et al. (2012).
31 Robert Jackson (1992), Stim Wilcox and Jackson (2002).
32 Andrew Whiten et al. (2005).
33 Edwin van Leeuwen and Daniel Haun (2013).

34 Susan Perry (2009); see also Marietta Dindo et al. (2009).

35 Elizabeth Lonsdorf et al. (2004).

36 Jenny Allen et al. (2013).

37 Erica van de Waal et al. (2013).

38 Nicolas Claidière et al. (2015).

39 Frans de Waal and Denise Johanowicz (1993).

40 Kristin Bonnie and Frans de Waal (2007).

41 Michio Nakamura et al. (2000).

42 Tetsuro Matsuzawa (1994), Noriko Inoue-Nakamura and Matsuzawa (1997).

43 Stuart Watson et al. (2015).

44 Tetsuro Matsuzawa et al. (2001), Frans de Waal (2001).

45 Konrad Lorenz (1952), p. 86.

46 Frans de Waal and Jennifer Pokorny (2008).

47 Frans de Waal and Peter Tyack (2003).

48 Stephanie King et al. (2013).

49 Laela Sayigh et al. (1999), Vincent Janik et al. (2006).

50 Jason Bruck (2013).

51 Stephanie King and Vincent Janik (2013).

CHAPTER 9: EVOLUTIONARY COGNITION

1 Marc Bekoff and Colin Allen (1997), p. 316.

2 Anthony Tramontin and Eliot Brenowitz (2000).

3 Jonathan Marks (2002), p. xvi.

4 David Hume (1985 [orig. 1739]), p. 226, with thanks to Gerald Massey.

5 "Study: Dolphins Not So Intelligent on Land," *Onion*, Feb. 15, 2006.

6 Jolyon Troscianko et al. (2012).

7 Donald Dewsbury (2000).

8 Frans de Waal and Sarah Brosnan (2006).

9 Frans de Waal and Pier Francesco Ferrari (2010).

BIBLIOGRAPHY

Adler, J. 2008. Thinking like a monkey. *Smithsonian Magazine,* January.

Aitchison, J. 2000. *The Seeds of Speech: Language Origin and Evolution*, Cambridge, UK: Cambridge University Press.

Alexander, M. G., and T. D. Fisher. 2003. Truth and consequences: Using the bogus pipeline to examine sex differences in self-reported sexuality. *Journal of Sex Research* 40:27–35.

Allen, B. 1997. The chimpanzee's tool. *Common Knowledge* 6:34–51.

Allen, J., M. Weinrich, W. Hoppitt, and L. Rendell. 2013. Network-based diffusion analysis reveals cultural transmission of lobtail feeding in humpback whales. *Science* 340:485–88.

Anderson, J. R., and G. G. Gallup. 2011. Which primates recognize themselves in mirrors? *Plos Biology* 9:e1001024.

Anderson, R. C., and J. A. Mather. 2010. It's all in the cues: Octopuses (*Enteroctopus dofleini*) learn to open jars. *Ferrantia* 59:8–13.

Anderson, R. C., J. A. Mather, M. Q. Monette, and S. R. M. Zimsen. 2010. Octopuses (*Enteroctopus dofleini*) recognize individual humans. *Journal of Applied Animal Welfare Science* 13:261–72.

Anderson, R. C., J. B. Wood, and R. A. Byrne. 2002. Octopus senescence: The beginning of the end. *Journal of Applied Animal Welfare Science* 5:275–83.

Aristotle. 1991. *History of Animals*, trans. D. M. Balme. Cambridge, MA: Harvard University Press.

Arnold, K., and K. Zuberbühler. 2008. Meaningful call combinations in a nonhuman primate. *Current Biology* 18:R202–3.

Auersperg, A. M. I., B. Szabo, A. M. P. Von Bayern, and A. Kacelnik. 2012. Spontaneous innovation in tool manufacture and use in a Goffin's cockatoo. *Current Biology* 22:R903–4.

Aureli, F., R. Cozzolinot, C. Cordischif, and S. Scucchi. 1992. Kin-oriented redirection among Japanese macaques: An expression of a revenge system? *Animal Behaviour* 44:283–91.

Azevedo, F. A. C., et al. 2009. Equal numbers of neuronal and nonneuronal cells make the human brain an isometrically scaled-up primate brain. *Journal of Comparative Neurology* 513:532–41.

Babb, S. J., and J. D. Crystal. 2006. Episodic-like memory in the rat. *Current Biology* 16:1317–21.

Ban, S. D., C. Boesch, and K. R. L. Janmaat. 2014. Taï chimpanzees anticipate revisiting high-valued fruit trees from further distances. *Animal Cognition* 17:1353–64.

Barton, R. A. 2012. Embodied cognitive evolution and the cerebellum. *Philosophical Transactions of the Royal Society B* 367:2097–107.

Bates, L. A., et al. 2007. Elephants classify human ethnic groups by odor and garment color. *Current Biology* 17:1938–42.

Baumeister, R. F. 2008. Free will in scientific psychology. *Perspectives on Psychological Science* 3:14–19.

Beach, F. A. 1950. The snark was a boojum. *American Psychologist* 5:115–24.

Beck, B. B. 1967. A study of problem-solving by gibbons. *Behaviour* 28: 95–109.

———. 1980. *Animal Tool Behavior: The Use and Manufacture of Tools by Animals.* New York: Garland STPM Press.

———. 1982. Chimpocentrism: Bias in cognitive ethology. *Journal of Human Evolution* 11:3–17.

Bekoff, M., and C. Allen. 1997. Cognitive ethology: Slayers, skeptics, and proponents. In *Anthropomorphism, Anecdotes, and Animals: The Emperor's New Clothes?* ed. R. W. Mitchell, N. Thompson, and L. Miles, 313–34. Albany: SUNY Press.

Bekoff, M., and P. W. Sherman. 2003. Reflections on animal selves. *Trends in Ecology and Evolution* 19:176–80.

Bekoff, M., C. Allen, and G. M. Burghardt, eds. 2002. *The Cognitive Animal: Empirical and Theoretical Perspectives on Animal Cognition.* Cambridge, MA: Bradford.

Beran, M. J. 2002. Maintenance of self-imposed delay of gratification by four chimpanzees (*Pan troglodytes*) and an orangutan (*Pongo pygmaeus*). *Journal of General Psychology* 129:49–66.

———. 2015. The comparative science of "self-control": What are we talking about? *Frontiers in Psychology* 6:51.

Berns, G. S. 2013. *How Dogs Love Us: A Neuroscientist and His Adopted Dog Decode the Canine Brain.* Boston: Houghton Mifflin.

Berns, G. S., A. Brooks, and M. Spivak. 2013. Replicability and heterogeneity of awake unrestrained canine fMRI responses. *Plos ONE* 8:e81698.

Bird, C. D., and N. J. Emery. 2009. Rooks use stones to raise the water level to reach a floating worm. *Current Biology* 19:1410–14.

Bischof-Köhler, D. 1991. The development of empathy in infants. In *Infant Development: Perspectives From German-Speaking Countries,* ed. M. Lamb and M. Keller, 245–73. Hillsdale, NJ: Erlbaum.

Bjorklund, D. F., J. M. Bering, and P. Ragan. 2000. A two-year longitudinal study of deferred imitation of object manipulation in a juvenile chimpanzee (*Pan troglodytes*) and orangutan (*Pongo pygmaeus*). *Developmental Psychobiology* 37:229–37.

Boesch, C. 2007. What makes us human? The challenge of cognitive cross-species comparison. *Journal of Comparative Psychology* 121:227–40.

Boesch, C., and H. Boesch-Achermann. 2000. *The Chimpanzees of the Taï Forest: Behavioural Ecology and Evolution.* Oxford: Oxford University Press.

Boesch, C., J. Head, and M. M. Robbins. 2009. Complex tool sets for honey extraction among chimpanzees in Loango National Park, Gabon. *Journal of Human Evolution* 56:560–69.

Bolhuis, J. J., and C. D. L. Wynne. 2009. Can evolution explain how minds work? *Nature* 458:832–33.

Bonnie, K. E., and F. B. M. de Waal. 2007. Copying without rewards: Socially influenced foraging decisions among brown capuchin monkeys. *Animal Cognition* 10: 283–92.

Bonnie, K. E., V. Horner, A. Whiten, and F. B. M. de Waal. 2006. Spread of arbitrary conventions among chimpanzees: A controlled experiment. *Proceedings of the Royal Society of London B* 274:367–72.

Bovet, D., and D. A. Washburn. 2003. Rhesus macaques categorize unknown conspecifics according to their dominance relations. *Journal of Comparative Psychology* 117:400–5.

Boyd, R. 2006. The puzzle of human sociality. *Science* 314:1555–56.

Boysen, S. T., and G. G. Berntson. 1989. Numerical competence in a chimpanzee (*Pan troglodytes*). *Journal of Comparative Psychology* 103:23–31.

———. 1995. Responses to quantity: Perceptual versus cognitive mechanisms in chimpanzees (*Pan troglodytes*). *Journal of Experimental Psychology: Animal Behavior Processes* 21:82–86.

Bramlett, J. L., B. M. Perdue, T. A. Evans, and M. J. Beran. 2012. Capuchin monkeys (*Cebus apella*) let lesser rewards pass them by to get better rewards. *Animal Cognition* 15:963–69.

Bräuer, J., et al. 2006. Making inferences about the location of hidden food: Social dog, causal ape. *Journal of Comparative Psychology* 120: 38–47.

Bräuer, J., and J. Call. 2015. Apes produce tools for future use. *American Journal of Primatology* 77:254–63.

Breland, K., and M. Breland. 1961. The misbehavior of organisms. *American Psychologist* 16:681–84.

Breuer, T., M. Ndoundou-Hockemba, and V. Fishlock. 2005. First observation of tool use in wild gorillas. *Plos Biology* 3:2041–43.

Brosnan, S. F., et al. 2010. Mechanisms underlying responses to inequitable outcomes in chimpanzees. *Animal Behaviour* 79:1229–37.

Brosnan, S. F., and F. B. M. de Waal. 2003. Monkeys reject unequal pay. *Nature* 425:297–99.

———. 2014. The evolution of responses to (un)fairness. *Science* 346:1251776.

Brosnan, S. F., C. Freeman, and F. B. M. de Waal. 2006. Partner's behavior, not reward distribution, determines success in an unequal cooperative task in capuchin monkeys. *American Journal of Primatology* 68:713–24.

Brown, C., M. P. Garwood, and J. E. Williamson. 2012. It pays to cheat: Tactical deception in a cephalopod social signalling system. *Biology Letters* 8:729–32.

Browning, R. 2006 [orig. 1896]. *The Poetical Works.* Whitefish, MT: Kessinger.

Bruck, J. N. 2013. Decades-long social memory in bottlenose dolphins. *Proceedings of the Royal Society B* 280: 20131726.

Bshary, R., and R. Noë. 2003. Biological markets: The ubiquitous influence of partner choice on the dynamics of cleaner fish-client reef fish interactions. In *Genetic and Cultural Evolution of Cooperation*, ed. P. Hammerstein, 167–84. Cambridge, MA: MIT Press.

Bshary, R., A. Hohner, K. Ait-El-Djoudi, and H. Fricke. 2006. Interspecific communicative and coordinated hunting between groupers and giant moray eels in the Red Sea. *Plos Biology* 4:e431.

Buchsbaum, R., M. Buchsbaum, J. Pearse, and V. Pearse. 1987. *Animals Without Backbones: An Introduction to the Invertebrates.* 3rd ed. Chicago: University of Chicago Press.

Buckley, J., et al. 2010. Biparental mucus feeding: A unique example of parental care in an Amazonian cichlid. *Journal of Experimental Biology* 213:3787–95.

Buckley, L. A., et al. 2011. Too hungry to learn? Hungry broiler breeders fail to learn a y-maze food quantity discrimination task. *Animal Welfare* 20: 469–81.

Bugnyar, T., and B. Heinrich. 2005. Ravens, *Corvus corax*, differentiate between knowledgeable and ignorant competitors. *Proceedings of the Royal Society of London B* 272:1641–46.

Burghardt, G. M. 1991. Cognitive ethology and critical anthropomorphism: A snake with two heads and hognose snakes that play dead. In *Cognitive Ethology: The Minds of Other Animals: Essays in Honor of Donald R. Griffin*, ed. C. A. Ristau, 53–90. Hillsdale, NJ: Lawrence Erlbaum Associates.

Burkhardt, R. W. 2005. *Patterns of Behavior: Konrad Lorenz, Niko Tinbergen, and the Founding of Ethology*. Chicago: University of Chicago Press.

Burrows, A. M., et al. 2006. Muscles of facial expression in the chimpanzee (*Pan troglodytes*): Descriptive, ecological and phylogenetic contexts. *Journal of Anatomy* 208:153–68.

Byrne, R. 1995. *The Thinking Ape: The Evolutionary Origins of Intelligence*. Oxford: Oxford University Press.

Byrne, R., and A. Whiten. 1988. *Machiavellian Intelligence*. Oxford: Oxford University Press.

Calcutt, S. E., et al. 2014. Captive chimpanzees share diminishing resources. *Behaviour* 151:1967–82.

Caldwell, C. C., and A. Whiten. 2002. Evolutionary perspectives on imitation: Is a comparative psychology of social learning possible? *Animal Cognition* 5:193–208.

Call, J. 2004. Inferences about the location of food in the great apes. *Journal of Comparative Psychology* 118:232–41.

———. 2006. Descartes' two errors: Reason and reflection in the great apes. In *Rational Animals*, ed. S. Hurley and M. Nudds, 219–234. Oxford: Oxford University Press.

Call, J., and M. Carpenter. 2001. Do apes and children know what they have seen? *Animal Cognition* 3:207–20.

Call, J., and M. Tomasello. 2008. Does the chimpanzee have a theory of mind? 30 Years Later. *Trends in Cognitive Sciences* 12:187–92.

Callaway, E. 2012. Alex the parrot's last experiment shows his mathematical genius. *Nature News Blog*, Feb. 20, http://bit.ly/1eYgqoD.

Calvin, W. H. 1982. Did throwing stones shape hominid brain evolution? *Ethology and Sociobiology* 3:115–24.

Candland, D. K. 1993. *Feral Children and Clever Animals: Reflections on Human Nature*. New York: Oxford University Press.

Cenami Spada, E., F. Aureli, P. Verbeek, and F. B. M. de Waal. 1995. The self as reference point: Can animals do without it? In *The Self in Infancy: Theory and Research*, ed. P. Rochat, 193–215. Amsterdam: Elsevier.

Chang, L., et al. 2015. Mirror-induced self-directed behaviors in rhesus monkeys after visual-somatosensory training. *Current Biology* 25:212–17.

Cheney, D. L., and R. M. Seyfarth. 1986. The recognition of social alliances by vervet monkeys. *Animal Behaviour* 34 (1986): 1722–31.

———. 1989. Redirected aggression and reconciliation among vervet monkeys, *Cercopithecus aethiops. Behaviour* 110: 258–75.

———. 1990. *How Monkeys See the World: Inside the Mind of Another Species.* Chicago: University of Chicago Press.

Claidière, N., et al. 2015. Selective and contagious prosocial resource donation in capuchin monkeys, chimpanzees and humans. *Scientific Reports* 5:7631.

Clayton, N. S., and A. Dickinson. 1998. Episodic-like memory during cache recovery by scrub jays. *Nature* 395:272–74.

Corballis, M. C. 2002. *From Hand to Mouth: The Origins of Language.* Princeton, NJ: Princeton University Press.

———. 2013. Mental time travel: A case for evolutionary continuity. *Trends in Cognitive Sciences* 17:5–6.

Corbey, R. 2005. *The Metaphysics of Apes: Negotiating the Animal-Human Boundary.* Cambridge: Cambridge University Press.

Correia, S. P. C., A. Dickinson, and N. S. Clayton. 2007. Western scrub-jays anticipate future needs independently of their current motivational state. *Current Biology* 17:856–61.

Courage, K. H. 2013. *Octopus! The Most Mysterious Creature in the Sea.* New York: Current.

Crawford, M. 1937. The cooperative solving of problems by young chimpanzees. *Comparative Psychology Monographs* 14:1–88.

Crockford, C., R. M. Wittig, R. Mundry, and K. Zuberbühler. 2012. Wild chimpanzees inform ignorant group members of danger. *Current Biology* 22:142–46.

Csányi, V. 2000. *If Dogs Could Talk: Exploring the Canine Mind.* New York: North Point Press.

Cullen, E. 1957. Adaptations in the kittiwake to cliff-nesting. *Ibis* 99:275–302.

Darwin, C. 1982 [orig. 1871]. *The Descent of Man, and Selection in Relation to Sex.* Princeton, NJ: Princeton University Press.

Davila Ross, M., M. J. Owren, and E. Zimmermann. 2009. Reconstructing the evolution of laughter in great apes and humans. *Current Biology* 19:1106–11.

de Groot, N. G., et al. 2010. AIDS-protective HLA-B★27/B★57 and chimpanzee MHC class I molecules target analogous conserved areas of HIV-1/SIVcpz. *Proceedings of the National Academy of Sciences, USA* 107:15175–80.

de Waal, F. B. M. 1991. Complementary methods and convergent evidence in the study of primate social cognition. *Behaviour* 118:297–320.

———. 1996. *Good Natured: The Origins of Right and Wrong in Humans and Other Animals.* Cambridge, MA: Harvard University Press.

———. 1997. *Bonobo: The Forgotten Ape.* Berkeley: University of California Press.

———. 1999. Anthropomorphism and anthropodenial: Consistency in our thinking about humans and other animals. *Philosophical Topics* 27:255–80.

———. 2000. Primates: A natural heritage of conflict resolution. *Science* 289: 586–90.

———. 2001. *The Ape and the Sushi Master: Cultural Reflections by a Primatologist.* New York: Basic Books.

———. 2003a. Darwin's legacy and the study of primate visual communication. In *Emotions Inside Out: 130 Years After Darwin's* "The Expression of the Emotions in Man and Animals," ed. P. Ekman, J. J. Campos, R. J. Davidson, and F. B. M. de Waal, 7–31. New York: New York Academy of Sciences.

———. 2003b. Silent invasion: Imanishi's primatology and cultural bias in science. *Animal Cognition* 6:293–99.

———. 2005. *Our Inner Ape.* New York: Riverhead.

———. 2007 [orig. 1982]. *Chimpanzee Politics: Power and Sex Among Apes.* Baltimore: Johns Hopkins University Press.

———. 2008. Putting the altruism back into altruism: The evolution of empathy. *Annual Review of Psychology* 59:279–300.

———. 2009a. *The Age of Empathy: Nature's Lessons for a Kinder Society.* New York: Harmony.

———. 2009b. Darwin's last laugh. *Nature* 460:175.

de Waal, F. B. M., and M. Berger. 2000. Payment for labour in monkeys. *Nature* 404:563.

de Waal, F. B. M., C. Boesch, V. Horner, and A. Whiten. 2008. Comparing children and apes not so simple. *Science* 319:569.

de Waal, F. B. M., and K. E. Bonnie. 2009. In tune with others: The social side of primate culture. In *The Question of Animal Culture*, ed. K. Laland and B. G. Galef, 19–39. Cambridge, MA: Harvard University Press.

de Waal, F. B. M., and S. F. Brosnan. 2006. Simple and complex reciprocity in primates. In *Cooperation in Primates and Humans: Mechanisms and Evolution*, ed. P. M. Kappeler and C. van Schaik, 85–105. Berlin: Springer.

de Waal, F. B. M., M. Dindo, C. A. Freeman, and M. Hall. 2005. The monkey in the mirror: Hardly a stranger. *Proceedings of the National Academy of Sciences USA* 102:11140–47.

de Waal, F. B. M., and P. F. Ferrari. 2010. Towards a bottom-up perspective on animal and human cognition. *Trends in Cognitive Sciences* 14:201–7.

de Waal, F. B. M., and D. L. Johanowicz. 1993. Modification of reconciliation behavior through social experience: An experiment with two macaque species. *Child Development* 64:897–908.

de Waal, F. B. M., and J. Pokorny. 2008. Faces and behinds: Chimpanzee sex perception. *Advanced Science Letters* 1:99–103.

de Waal, F. B. M., and P. L. Tyack, eds. 2003. *Animal Social Complexity: Intelligence, Culture, and Individualized Societies.* Cambridge, MA: Harvard University Press.

de Waal, F. B. M., and J. van Hooff. 1981. Side-directed communication and agonistic interactions in chimpanzees. *Behaviour* 77:164–98.

Dewsbury, D. A. 2000. Comparative cognition in the 1930s. *Psychonomic Bulletin and Review* 7:267–83.

———. 2006. *Monkey Farm: A History of the Yerkes Laboratories of Primate Biology, Orange Park, Florida, 1930–1965.* Lewisburg, PA: Bucknell University Press.

Dindo, M., A. Whiten, and F. B. M. de Waal. 2009. In-group conformity sustains different foraging traditions in capuchin monkeys (*Cebus apella*). *Plos ONE* 4:e7858.

Dinets, V., J. C. Brueggen, and J. D. Brueggen. 2013. Crocodilians use tools for hunting. *Ethology Ecology and Evolution* 27:74–78.

Dingfelder, S. D. 2007. Can rats reminisce? *Monitor on Psychology* 38:26.

Domjan, M., and B. G. Galef. 1983. Biological constraints on instrumental and classical conditioning: Retrospect and prospect. *Animal Learning and Behavior* 11:151–61.

Ducheminsky, N., P. Henzi, and L. Barrett. 2014. Responses of vervet monkeys in large troops to terrestrial and aerial predator alarm calls. *Behavioral Ecology* 25:1474–84.

Dunbar, R. 1998a. *Grooming, Gossip, and the Evolution of Language.* Cambridge, MA: Harvard University Press.

———. 1998b. The social brain hypothesis. *Evolutionary Anthropology* 6:178–90.

Emery, N. J., and N. S. Clayton. 2001. Effects of experience and social context on prospective caching strategies by scrub jays. *Nature* 414:443–46.

———. 2004. The mentality of crows: Convergent evolution of intelligence in corvids and apes. *Science* 306:1903–7.

Epstein, R. 1987. The spontaneous interconnection of four repertoires of behavior in a pigeon. *Journal of Comparative Psychology* 101:197–201.

Epstein, R., R. P. Lanza, and B. F. Skinner. 1981. "Self-awareness" in the pigeon. *Science* 212:695–96.

Evans, T. A., and M. J. Beran. 2007. Chimpanzees use self-distraction to cope with impulsivity. *Biology Letters* 3:599–602.

Falk, J. L. 1958. The grooming behavior of the chimpanzee as a reinforcer. *Journal of the Experimental Analysis of Behavior* 1:83–85.

Fehr, E., and U. Fischbacher. 2003. The nature of human altruism. *Nature* 425:785–91.

Ferris, C. F., et al. 2001. Functional imaging of brain activity in conscious monkeys responding to sexually arousing cues. *Neuroreport* 12:2231–36.

Finn, J. K., T. Tregenza, and M. D. Norman. 2009. Defensive tool use in a coconut-carrying octopus. *Current Biology* 19:R1069–70.

Fodor, J. 1975. *The Language of Thought.* New York: Crowell.

Foerder, P., et al. 2011. Insightful problem solving in an Asian elephant. *Plos ONE* 6(8):e23251.

Foote, A. L., and J. D. Crystal. 2007. Metacognition in the rat. *Current Biology* 17:551–55.

Foster, M. W., et al. 2009. Alpha male chimpanzee grooming patterns: Implications for dominance "style." *American Journal of Primatology* 71:136–44.

Fragaszy, D. M., E. Visalberghi, and L. M. Fedigan. 2004. *The Complete Capuchin: The Biology of the Genus* Cebus. Cambridge: Cambridge University Press.

Frankfurt, H. G. 1971. Freedom of the will and the concept of a person. *Journal of Philosophy* 68:5–20.

Fuhrmann, D., A. Ravignani, S. Marshall-Pescini, and A. Whiten. 2014. Synchrony and motor mimicking in chimpanzee observational learning. *Scientific Reports* 4:5283.

Gácsi, M., et al. 2009. Explaining dog wolf differences in utilizing human pointing gestures: Selection for synergistic shifts in the development of some social skills. *Plos ONE* 4:e6584.

Galef, B. G. 1990. The question of animal culture. *Human Nature* 3:157–78.

Gallup, G. G. 1970. Chimpanzees: Self-recognition. *Science* 167:86–87.

Garcia, J., D. J. Kimeldorf, and R. A. Koelling. 1955. Conditioned aversion to saccharin resulting from exposure to gamma radiation. *Science* 122:157–58.

Gardner, R. A., M. H. Scheel, and H. L. Shaw. 2011. Pygmalion in the laboratory. *American Journal of Psychology* 124:455–61.

Garstang, M., et al. 2014. Response of African elephants (*Loxodonta africana*) to seasonal changes in rainfall. *Plos ONE* 9:e108736.

Gaulin, S. J. C., and R. W. Fitzgerald. 1989. Sexual selection for spatial-learning ability. *Animal Behaviour* 37:322–31.

Geissmann, T., and M. Orgeldinger. 2000. The relationship between duet songs and pair bonds in siamangs, *Hylobates syndactylus*. *Animal Behaviour* 60: 805–9.

Goodall, J. 1967. *My Friends the Wild Chimpanzees*. Washington, DC: National Geographic Society.

———. 1971. *In the Shadow of Man*. Boston: Houghton Mifflin.

———. 1986. *The Chimpanzees of Gombe: Patterns of Behavior*. Cambridge, MA: Belknap.

Gould, J. L., and C. G. Gould. 1999. *The Animal Mind*. New York: W. H. Freeman.

Gouzoules, S., H. Gouzoules, and P. Marler. 1984. Rhesus monkey (*Macaca mulatta*) screams: Representational signaling in the recruitment of agonistic aid. *Animal Behaviour* 32:182–93.

Griffin, D. R. 1976. *The Question of Animal Awareness: Evolutionary Continuity of Mental Experience*. New York: Rockefeller University Press.

———. 2001. Return to the magic well: Echolocation behavior of bats and responses of insect prey. *Bioscience* 51:555–56.

Gruber, T., Z. Clay, and K. Zuberbühler. 2010. A comparison of bonobo and chimpanzee tool use: Evidence for a female bias in the Pan lineage. *Animal Behaviour* 80:1023–33.

Guldberg, H. 2010. *Just Another Ape?* Exeter, UK: Imprint Academic.

Gumert, M. D., M. Kluck, and S. Malaivijitnond. 2009. The physical characteristics and usage patterns of stone axe and pounding hammers used by long-tailed macaques in the Andaman Sea region of Thailand. *American Journal of Primatology* 71:594–608.

Günther, M. M., and C. Boesch. 1993. Energetic costs of nut-cracking behaviour in wild chimpanzees. In *Hands of Primates,* ed. H. Preuschoft and D. J. Chivers, 109–29. Vienna: Springer.

Gupta, A. S., M. A. A. van der Meer, D. S. Touretzky, and A. D. Redish. 2010. Hippocampal replay is not a simple function of experience. *Neuron* 65:695–705.

Guthrie, E. R., and G. P. Horton. 1946. *Cats in a Puzzle Box*. New York: Rinehart.

Hall, K., et al. 2014. Using cross correlations to investigate how chimpanzees use conspecific gaze cues to extract and exploit information in a foraging competition. *American Journal of Primatology* 76:932–41.

Hamilton, G. 2012. Crows can distinguish faces in a crowd. National Wildlife Federation, Nov. 7, http://bit.ly/1IqkWaN.

Hampton, R. R. 2001. Rhesus monkeys know when they remember. *Proceedings of the National Academy of Sciences USA* 98:5359–62.

Hampton, R. R., A. Zivin, and E. A. Murray. 2004. Rhesus monkeys (*Macaca mulatta*) discriminate between knowing and not knowing and collect information as needed before acting. *Animal Cognition* 7:239–54.

Hanlon, R. T. 2007. Cephalopod dynamic camouflage. *Current Biology* 17: R400–4.

———. 2013. Camouflaged octopus makes marine biologist scream bloody murder (video). *Discover*, Sept. 13, http://bit.ly/1RScdid.

Hanlon, R. T., and J. B. Messenger. 1996. *Cephalopod Behaviour*. Cambridge: Cambridge University Press.

Hanlon, R. T., J. W. Forsythe, and D. E. Joneschild. 1999. Crypsis, conspicuousness, mimicry and polyphenism as antipredator defences of foraging octopuses on indo-pacific coral reefs, with a method of quantifying crypsis from video tapes. *Biological Journal of the Linnean Society* 66:1–22.

Hanus, D., N. Mendes, C. Tennie, and J. Call. 2011. Comparing the performances of apes (*Gorilla gorilla, Pan troglodytes, Pongo pygmaeus*) and human children (*Homo sapiens*) in the floating peanut task. *PLoS ONE* 6:e19555.

Hare, B., M. Brown, C. Williamson, and M. Tomasello. 2002. The domestication of social cognition in dogs. *Science* 298:1634–36.

Hare, B., J. Call, and M. Tomasello 2001. Do chimpanzees know what conspecifics know? *Animal Behaviour* 61:139–51.

Hare, B., and M. Tomasello. 2005. Human-like social skills in dogs? *Trends in Cognitive Sciences* 9:440–45.

Hare, B., and V. Woods. 2013. *The Genius of Dogs: How Dogs Are Smarter Than You Think*. New York: Dutton.

Harlow, H. F. 1953. Mice, monkeys, men, and motives. *Psychological Review* 60: 23–32.

Hattori, Y., F. Kano, and M. Tomonaga. 2010. Differential sensitivity to conspecific and allospecific cues in chimpanzees and humans: A comparative eye-tracking study. *Biology Letters* 6:610–13.

Hattori, Y., K. Leimgruber, K. Fujita, and F. B. M. de Waal. 2012. Food-related tolerance in capuchin monkeys (*Cebus apella*) varies with knowledge of the partner's previous food-consumption. *Behaviour* 149:171–85.

Heisenberg, W. 1958. *Physics and Philosophy: The Revolution in Modern Science*. London: Allen and Unwin.

Herculano-Houzel, S. 2009. The human brain in numbers: A linearly scaled-up primate brain. *Frontiers in Human Neuroscience* 3 (2009): 1–11.

————. 2011. Brains matter, bodies maybe not: The case for examining neuron numbers irrespective of body size. *Annals of the New York Academy of Sciences* 1225:191–99.

Herculano-Houzel, S., et al. 2014. The elephant brain in numbers. *Neuroanatomy* 8:10.3389/fnana.2014.00046.

Herrmann, E., et al. 2007. Humans have evolved specialized skills of social cognition: The cultural intelligence hypothesis. *Science* 317:1360–66.

Herrmann, E., V. Wobber, and J. Call. 2008. Great apes' (*Pan troglodytes, P. paniscus, Gorilla gorilla, Pongo pygmaeus*) understanding of tool functional properties after limited experience. *Journal of Comparative Psychology* 122:220–30.

Heyes, C. 1995. Self-recognition in mirrors: Further reflections create a hall of mirrors. *Animal Behaviour* 50: 1533–42.

Hillemann, F., T. Bugnyar, K. Kotrschal, and C. A. F. Wascher. 2014. Waiting for better, not for more: Corvids respond to quality in two delay maintenance tasks. *Animal Behaviour* 90: 1–10.

Hirata, S., K. Watanabe, and M. Kawai. 2001. "Sweet-potato washing" revisited. In *Primate Origins of Human Cognition and Behavior*, ed. T. Matsuzawa, 487–508. Tokyo: Springer.

Hobaiter, C., and R. Byrne. 2014. The meanings of chimpanzee gestures. *Current Biology* 24:1596–600.

Hodos, W., and C. B. G. Campbell. 1969. *Scala naturae:* Why there is no theory in comparative psychology. *Psychological Review* 76:337–50.

Hopper, L. M., S. P. Lambeth, S. J. Schapiro, and A. Whiten. 2008. Observational learning in chimpanzees and children studied through "ghost" conditions. *Proceedings of the Royal Society of London B* 275:835–40.

Horner, V., et al. 2010. Prestige affects cultural learning in chimpanzees. *Plos ONE* 5:e10625.

Horner, V., D. J. Carter, M. Suchak, and F. B. M. de Waal. 2011. Spontaneous prosocial choice by chimpanzees. *Proceedings of the Academy of Sciences, USA* 108:13847–51.

Horner, V., and F. B. M. de Waal. 2009. Controlled studies of chimpanzee cultural transmission. *Progress in Brain Research* 178:3–15.

Horner, V., A. Whiten, E. Flynn, and F. B. M. de Waal. 2006. Faithful replication of foraging techniques along cultural transmission chains by chimpanzees and children. *Proceedings of the National Academy of Sciences USA* 103:13878–83.

Horowitz, A. 2010. *Inside of a Dog: What Dogs See, Smell, and Know.* New York: Scribner.

Hostetter, A. B., M. Cantero, and W. D. Hopkins. 2001. Differential use of vocal and gestural communication by chimpanzees (*Pan troglodytes*) in response to the attentional status of a human (*Homo sapiens*). *Journal of Comparative Psychology* 115:337–43.

Howell, T. J., S. Toukhsati, R. Conduit, and P. Bennett. 2013. The perceptions of dog intelligence and cognitive skills (PoDIaCS) survey. *Journal of Veterinary Behavior: Clinical Applications and Research* 8:418–24.

Huffman, M. A. 1996. Acquisition of innovative cultural behaviors in nonhuman primates: A case study of stone handling, a socially transmitted behavior in Japanese macaques. In *Social Learning in Animals: The Roots of Culture,* ed. C. M. Heyes and B. Galef, 267–89. San Diego: Academic Press.

Hume, D. 1985 [orig. 1739]. *A Treatise of Human Nature.* Harmondsworth, UK: Penguin.

Hunt, G. R. 1996. The manufacture and use of hook tools by New Caledonian crows. *Nature* 379:249–51.

Hunt, G. R., et al. 2007. Innovative pandanus-folding by New Caledonian crows. *Australian Journal of Zoology* 55:291–98.

Hunt, G. R., and R. D. Gray. 2004. The crafting of hook tools by wild New Caledonian crows. *Proceedings of the Royal Society of London* B 271:S88–S90.

Hurley, S., and M. Nudds. 2006. *Rational Animals?* Oxford: Oxford University Press.

Imanishi, K. *Man.* 1952. Tokyo: Mainichi-Shinbunsha.

Inman, A., and S. J. Shettleworth. 1999. Detecting metamemory in nonverbal subjects: A test with pigeons. *Journal of Experimental Psychology: Animal Behavior Processes* 25:389–95.

Inoue, S., and T. Matsuzawa. 2007. Working memory of numerals in chimpanzees. *Current Biology* 17:R1004–R1005.

Inoue-Nakamura, N., and T. Matsuzawa. 1997. Development of stone tool use by wild chimpanzees. *Journal of Comparative Psychology* 111:159–73.

Itani, J., and A. Nishimura. 1973. The study of infrahuman culture in Japan: A review. In *Precultural Primate Behavior,* ed. E. Menzel, 26–50. Basel: Karger.

Jabr, F. 2014. The science is in: Elephants are even smarter than we realized. *Scientific American,* Feb. 26.

Jackson, R. R. 1992. Eight-legged tricksters. *Bioscience* 42:590–98 .

Jacobs, L. F., and E. R. Liman. 1991. Grey squirrels remember the locations of buried nuts. *Animal Behaviour* 41:103–10.

Janik, V. M., L. S. Sayigh, and R. S. Wells. 2006. Signature whistle contour shape conveys identity information to bottlenose dolphins. *Proceedings of the National Academy of Sciences USA* 103:8293–97.

Janmaat, K. R. L., L. Polansky, S. D. Ban, and C. Boesch. 2014. Wild chimpanzees plan their breakfast time, type, and location. *Proceedings of the National Academy of Sciences USA* 111:16343–48.

Jelbert, S. A., et al. 2014. Using the Aesop's fable paradigm to investigate causal understanding of water displacement by New Caledonian crows. *Plos ONE* 9:e92895.

Jorgensen, M. J., S. J. Suomi, and W. D. Hopkins. 1995. Using a computerized testing system to investigate the preconceptual self in nonhuman primates and humans. In *The Self in Infancy: Theory and Research,* ed. P. Rochat, 243–256. Amsterdam: Elsevier.

Judge, P. G. 1991. Dyadic and triadic reconciliation in pigtail macaques (*Macaca nemestrina*). *American Journal of Primatology* 23:225–37.

Judge, P. G., and S. H. Mullen. 2005. Quadratic postconflict affiliation among bystanders in a hamadryas baboon group. *Animal Behaviour* 69:1345–55.

Kagan, J. 2000. Human morality is distinctive. *Journal of Consciousness Studies* 7:46–48.

———. 2004. The uniquely human in human nature. *Daedalus* 133:77–88.

Kaminski, J., J. Call, and J. Fischer. 2004. Word learning in a domestic dog: evidence for fast mapping. *Science* 304:1682–83.

Kendal, R., et al. 2015. Chimpanzees copy dominant and knowledgeable individuals: Implications for cultural diversity. *Evolution and Human Behavior* 36:65–72.

Kinani, J.-F., and D. Zimmerman. 2015. Tool use for food acquisition in a wild mountain gorilla (*Gorilla beringei beringei*). *American Journal of Primatology* 77:353–57.

King, S. L., and V. M. Janik. 2013. Bottlenose dolphins can use learned vocal labels to address each other. *Proceedings of the National Academy of Sciences USA* 110: 13216–21.

King, S. L., et al. 2013. Vocal copying of individually distinctive signature whistles in bottlenose dolphins. *Proceedings of the Royal Society* B 280: 20130053.

Kitcher, P. 2006. Ethics and evolution: How to get here from there. In *Primates and Philosophers: How Morality Evolved,* ed. S. Macedo and J. Ober, 120–39. Princeton, NJ: Princeton University Press.

Koepke, A. E., S. L. Gray, and I. M. Pepperberg. 2015. Delayed gratification: A grey parrot (*Psittacus erithacus*) will wait for a better reward. *Journal of Comparative Psychology*. In press.

Köhler, W. 1925. *The Mentality of Apes*. New York: Vintage.

Koyama, N. F. 2001. The long-term effects of reconciliation in Japanese macaques (*Macaca fuscata*). *Ethology* 107:975–87.

Koyama, N. F., C. Caws, and F. Aureli. 2006. Interchange of grooming and agonistic support in chimpanzees. *International Journal of Primatology* 27:1293–309.

Kruuk, H. 2003. *Niko's Nature: The Life of Niko Tinbergen and His Science of Animal Behaviour*. Oxford: Oxford University Press.

Kummer, H. 1971. *Primate Societies: Group Techniques of Ecological Adaptions*. Chicago: Aldine.

———. 1995. *In Quest of the Sacred Baboon: A Scientist's Journey*. Princeton, NJ: Princeton University Press.

Kummer, H., V. Dasser, and P. Hoyningen-Huene. 1990. Exploring primate social cognition: Some critical remarks. *Behaviour* 112:84–98.

Kuroshima, H., et al. 2003. A capuchin monkey recognizes when people do and do not know the location of food. *Animal Cognition* 6:283–91.

Ladygina-Kohts, N. 2002 [orig. 1935]. *Infant Chimpanzee and Human Child: A Classic 1935 Comparative Study of Ape Emotions and Intelligence*, ed. F. B. M. de Waal. Oxford: Oxford University Press.

Langergraber, K. E., J. C. Mitani, and L. Vigilant. 2007. The limited impact of kinship on cooperation in wild chimpanzees. *Proceedings of the Academy of Sciences USA* 104:7786–90.

Lanner, R. M. 1996. *Made for Each Other: A Symbiosis of Birds and Pines*. New York: Oxford University Press.

Leavens, D. A., F. Aureli, W. D. Hopkins, and C. W. Hyatt. 2001. Effects of cognitive challenge on self-directed behaviors by chimpanzees (*Pan troglodytes*). *American Journal of Primatology* 55:1–14.

Leavens, D., W. D. Hopkins, and K. A. Bard. 1996. Indexical and referential pointing in chimpanzees (*Pan troglodytes*). *Journal of Comparative Psychology* 110 (1996): 346–53.

Lehrman, D. 1953. A critique of Konrad Lorenz's theory of instinctive behavior. *Quarterly Review of Biology* 28:337–63.

Lethmate, J. 1982. Tool-using skills of orangutans. *Journal of Human Evolution* 11:49–50.

Lethmate, J., and G. Dücker. 1973. Untersuchungen zum Selbsterkennen im

Spiegel bei Orang-Utans und einigen anderen Affenarten. *Zeitschrift für Tierpsychologie* 33:248–69.

Liebal, K., B. M. Waller, A. M. Burrows, and K. E. Slocombe. 2013. *Primate Communication: A Multimodal Approach.* Cambridge: Cambridge University Press.

Limongelli, L., S. Boysen, and E. Visalberghi. 1995. Comprehension of cause-effect relations in a tool-using task by chimpanzees (*Pan troglodytes*). *Journal of Comparative Psychology* 109:18–26.

Lindauer, M. 1987. Introduction. In *Neurobiology and Behavior of Honeybees*, ed. R. Menzel and A. Mercer, 1–6. Berlin: Springer.

Lonsdorf, E. V., L. E. Eberly, and A. E. Pusey. 2004. Sex differences in learning in chimpanzees. *Nature* 428:715–16.

Lorenz, K. Z. 1941. Vergleichende Bewegungsstudien an Anatinen. *Journal für Ornithologie* 89 (1941): 194–294.

———. 1952. *King Solomon's Ring.* London: Methuen, 1952.

———. 1981. *The Foundations of Ethology.* New York: Simon and Schuster.

Malcolm, N. 1973. Thoughtless brutes. *Proceedings and Addresses of the American Philosophical Association* 46:5–20.

Marais, E. 1969. *The Soul of the Ape.* New York: Atheneum.

Marks, J. 2002. *What It Means to Be 98% Chimpanzee: Apes, People, and Their Genes.* Berkeley: University of California Press.

Martin, C. F., et al. 2014. Chimpanzee choice rates in competitive games match equilibrium game theory predictions. *Scientfic Reports* 4:5182.

Martin-Ordas, G., D. Berntsen, and J. Call. 2013. Memory for distant past events in chimpanzees and orangutans. *Current Biology* 23:1438–41.

Martin-Ordas, G., J. Call, and F. Colmenares. 2008. Tubes, tables and traps: Great apes solve two functionally equivalent trap tasks but show no evidence of transfer across tasks. *Animal Cognition* 11:423–30.

Marzluff, J. M., et al. 2010. Lasting recognition of threatening people by wild American crows. *Animal Behaviour* 79:699–707.

Marzluff, J. M., and T. Angell. 2005. *In the Company of Crows and Ravens.* New Haven, CT: Yale University Press.

Marzluff, J. M., R. Miyaoka, S. Minoshima, and D. J. Cross. 2012. Brain imaging reveals neuronal circuitry underlying the crow's perception of human faces. *Proceedings of the National Academy of Sciences USA* 109:15912–17.

Mason, W. A. 1976. Environmental models and mental modes: Representational processes in the great apes and man. *American Psychologist* 31:284–94.

Massen, J. J. M., A. Pašukonis, J. Schmidt, and T. Bugnyar. 2014. Ravens notice dominance reversals among conspecifics within and outside their social group. *Nature Communications* 5:3679.

Massen, J. J. M., G. Szipl, M. Spreafico, and T. Bugnyar. 2014. Ravens intervene in others' bonding attempts. *Current Biology* 24:2733–36.

Mather, J. A., and R. C. Anderson. 1999. Exploration, play, and habituation in octopuses (*Octopus dofleini*). *Journal of Comparative Psychology* 113:333–38.

Mather, J. A., R. C. Anderson, and J. B. Wood. 2010. *Octopus: The Ocean's Intelligent Invertebrate*. Portland, OR: Timber Press.

Matsuzawa, T. 1994. Field experiments on use of stone tools by chimpanzees in the wild. In *Chimpanzee Cultures,* ed. R. W. Wrangham, W. C. McGrew, F. B. M. de Waal, and P. Heltne, 351–70. Cambridge, MA: Harvard University Press.

———. 2009. Symbolic representation of number in chimpanzees. *Current Opinion in Neurobiology* 19:92–98.

Matsuzawa, T., et al. 2001. Emergence of culture in wild chimpanzees: education by master-apprenticeship. In *Primate Origins of Human Cognition and Behavior*, ed. T. Matsuzawa, 557–74. New York: Springer.

Mayr, E. 1982. *The Growth of Biological Thought*. Cambridge, MA: Harvard University Press.

McComb, K., et al. 2011. Leadership in elephants: The adaptive value of age. *Proceedings of the Royal Society B* 274:2943–49.

McComb, K., G. Shannon, K. N. Sayialel, and C. Moss. 2014. Elephants can determine ethnicity, gender and age from acoustic cues in human voices. *Proceedings of the National Academy of Sciences USA* 111:5433–38.

McGrew, W. C. 2010. Chimpanzee technology. *Science* 328:579–80.

———. 2013. Is primate tool use special? Chimpanzee and New Caledonian crow compared. *Philosophical Transactions of the Royal Society B* 368:20120422.

McGrew, W. C., and C. E. G. Tutin. 1978. Evidence for a social custom in wild chimpanzees? *Man* 13:243–51.

Melis, A. P., B. Hare, and M. Tomasello. 2006a. Chimpanzees recruit the best collaborators. *Science* 311:1297–300.

———. 2006b. Engineering cooperation in chimpanzees: Tolerance constraints on cooperation. *Animal Behaviour* 72:275–86.

Mendes, N., D. Hanus, and J. Call. 2007. Raising the level: Orangutans use water as a tool. *Biology Letters* 3:453–55.

Mendres, K. A., and F. B. M. de Waal. 2000. Capuchins do cooperate: The advantage of an intuitive task. *Animal Behaviour* 60: 523–29.

Menzel, E. W. 1972. Spontaneous invention of ladders in a group of young chimpanzees. *Folia primatologica* 17:87–106.

———. 1974. A group of young chimpanzees in a one-acre field. In *Behavior of Non-Human Primates*, ed. A. M. Schrier and F. Stollnitz, 5:83–153. New York: Academic Press.

Mercader, J., et al. 2007. 4,300-year-old chimpanzee sites and the origins of percussive stone technology. *Proceedings of the National Academy of Sciences USA* 104:3043–48.

Miklósi, Á., et al. 2003. A simple reason for a big difference: Wolves do not look back at humans, but dogs do. *Current Biology* 13:763–66.

Mischel, W., and E. B. Ebbesen. 1970. Attention in delay of gratification. *Journal of Personality and Social Psychology* 16:329–37.

Mischel, W., E. B. Ebbesen, and A. R. Zeiss. 1972. Cognitive and attentional mechanisms in delay of gratification. *Journal of Personality and Social Psychology* 21:204–18.

Moore, B. R. 1973. The role of directed pavlovian responding in simple instrumental learning in the pigeon. In *Constraints on Learning,* ed. R. A. Hinde and J. S. Hinde, 159–87. London: Academic Press.

———. 1992. Avian movement imitation and a new form of mimicry: Tracing the evoluting of a complex form of learning. *Behaviour* 122:231–63.

———. 2004. The evolution of learning. *Biological Review* 79:301–35.

Moore, B. R., and S. Stuttard. 1979. Dr. Guthrie and *Felis domesticus* or: Tripping over the cat. *Science* 205:1031–33.

Morell, V. 2013. *Animal Wise: The Thoughts and Emotions of Our Fellow Creatures.* New York: Crown.

Morgan, C. L. 1894. *An Introduction to Comparative Psychology.* London: Scott.

———. 1903. *An Introduction to Comparative Psychology,* new ed. London: Scott.

Morris, D. 2010. Retrospective: Beginnings. In *Tinbergen's Legacy in Behaviour: Sixty Years of Landmark Stickleback Papers*, ed. F. Von Hippel, 49–53. Leiden, Netherlands: Brill.

Morris, R., and D. Morris. 1966. *Men and Apes.* New York: McGraw-Hill.

Mulcahy, N. J., and J. Call. 2006. Apes save tools for future use. *Science* 312:1038–40.

Nagasawa, M., et al. 2015. Oxytocin-gaze positive loop and the co-evolution of human-dog bonds. *Science* 348:333–36.

Nagel, T. 1974. What is it like to be a bat? *Philosophical Review* 83:435–50.

Nakamura, M., W. C. McGrew, L. F. Marchant, and T. Nishida. 2000. Social scratch: Another custom in wild chimpanzees? *Primates* 41:237–48.

Neisser, U. 1967. *Cognitive Psychology.* Englewood Cliffs, NJ: Prentice-Hall.

Nielsen, R., et al. 2005. A scan for positively selected genes in the genomes of humans and chimpanzees. *Plos Biology* 3:976–85.

Nishida, T. 1983. Alpha status and agonistic alliances in wild chimpanzees. *Primates* 24:318–36.

Nishida, T., et al. 1992. Meat-sharing as a coalition strategy by an alpha male chimpanzee? In *Topics of Primatology*, ed. T. Nishida, 159–74. Tokyo: Tokyo Press.

Nishida, T., and K. Hosaka. 1996. Coalition strategies among adult male chimpanzees of the Mahale Mountains, Tanzania. In *Great Ape Societies* ed. W. C. McGrew, L. F. Marchant, and T. Nishida, 114–34. Cambridge: Cambridge University Press.

O'Connell, C. 2015. *Elephant Don: The Politics of a Pachyderm Posse.* Chicago: University of Chicago Press.

Ostojić, L., R. C. Shaw, L. G. Cheke, and N. S. Clayton. 2013. Evidence suggesting that desire-state attribution may govern food sharing in Eurasian jays. *Proceedings of the National Academy of Sciences USA* 110: 4123–28.

Osvath, M. 2009. Spontaneous planning for stone throwing by a male chimpanzee. *Current Biology* 19:R191–92.

Osvath, M., and G. Martin-Ordas. 2014. The future of future-oriented cognition in non-humans: Theory and the empirical case of the great apes. *Philosophical Transactions of the Royal Society B* 369:20130486.

Osvath, M., and H. Osvath. 2008. Chimpanzee (*Pan troglodytes*) and orangutan (*Pongo abelii*) forethought: Self-control and pre-experience in the face of future tool use. *Animal Cognition* 11:661–74.

Ottoni, E. B., and M. Mannu. 2001. Semifree-ranging tufted capuchins (*Cebus apella*) spontaneously use tools to crack open nuts. *International Journal of Primatology* 22:347–58.

Overduin-de Vries, A. M., B. M. Spruijt, and E. H. M. Sterck. 2013. Long-tailed macaques (*Macaca fascicularis*) understand what conspecifics can see in a competitive situation. *Animal Cognition* 17:77–84.

Parr, L., and F. B. M. de Waal. 1999. Visual kin recognition in chimpanzees. *Nature* 399:647-48.

Parvizi, J. 2009. Corticocentric myopia: Old bias in new cognitive sciences. *Trends in Cognitive Sciences* 13:354–59.

Paxton, R., et al. 2010. Rhesus monkeys rapidly learn to select dominant individuals in videos of artificial social interactions between unfamiliar conspecifics. *Journal of Comparative Psychology* 124:395–401.

Pearce, J. M. 2008. *Animal Learning and Cognition: An Introduction*, 3rd ed. East Sussex, UK: Psychology Press.

Penn, D. C., and D. J. Povinelli. 2007. On the lack of evidence that non-human animals possess anything remotely resembling a "theory of mind." *Philosophical Transactions of the Royal Society B* 362:731–44.

Pepperberg, I. M. 1999. *The Alex Studies: Cognitive and Communicative Abilities of Grey Parrots*. Cambridge, MA: Harvard University Press.

———. 2008. *Alex and Me*. New York: Collins.

———. 2012. Further evidence for addition and numerical competence by a grey parrot (*Psittacus erithacus*). *Animal Cognition* 15:711–17.

Perdue, B. M., R. J. Snyder, Z. Zhihe, M. J. Marr, and T. L. Maple. 2011. Sex differences in spatial ability: A test of the range size hypothesis in the order Carnivora. *Biology Letters* 7:380–83.

Perry, S. 2008. *Manipulative Monkeys: The Capuchins of Lomas Barbudal*. Cambridge, MA: Harvard University Press.

———. 2009. Conformism in the food processing techniques of white-faced capuchin monkeys (*Cebus capucinus*). *Animal Cognition* 12:705–16.

Perry, S., H. Clark Barrett, and J. H. Manson. 2004. White-faced capuchin monkeys show triadic awareness in their choice of allies. *Animal Behaviour* 67:165–70.

Pfenning, A. R., et al. 2014. Convergent transcriptional specializations in the brains of humans and song-learning birds. *Science* 346:1256846.

Pfungst, O. 1911. *Clever Hans (The Horse of Mr. von Osten): A Contribution to Experimental Animal and Human Psychology*. New York: Henry Holt.

Plotnik, J. M., et al. 2014. Thinking with their trunks: Elephants use smell but not sound to locate food and exclude nonrewarding alternatives. *Animal Behaviour* 88:91–98.

Plotnik, J. M., F. B. M. de Waal, and D. Reiss. 2006. Self-recognition in an Asian elephant. *Proceedings of the National Academy of Sciences USA* 103:17053–57.

Plotnik, J. M., R. C. Lair, W. Suphachoksakun, and F. B. M. de Waal. 2011. Elephants know when they need a helping trunk in a cooperative task. *Proceedings of the Academy of Sciences USA* 108:516–21.

Pokorny, J., and F. B. M. de Waal. 2009. Monkeys recognize the faces of group mates in photographs. *Proceedings of the National Academy of Sciences USA* 106:21539–43.

Pollick, A. S., and F. B. M. de Waal. 2007. Ape gestures and language evolution. *Proceedings of the National Academy of Sciences USA* 104:8184–89.

Povinelli, D. J. 1987. Monkeys, apes, mirrors and minds: The evolution of self-awareness in primates. *Human Evolution* 2:493–509.

———. 1989. Failure to find self-recognition in Asian elephants (*Elephas maximus*) in contrast to their use of mirror cues to discover hidden food. *Journal of Comparative Psychology* 103:122–31.

———. 1998. Can animals empathize? *Scientific American Presents: Exploring Intelligence* 67:72–75.

———. 2000. *Folk Physics for Apes: The Chimpanzee's Theory of How the World Works*. Oxford: Oxford University Press.

Povinelli, D. J., et al. 1997. Chimpanzees recognize themselves in mirrors. *Animal Behaviour* 53:1083–88.

Premack, D. 2007. Human and animal cognition: Continuity and discontinuity. *Proceedings of the National Academy of Sciences USA* 104:13861–67.

———. 2010. Why humans are unique: Three theories. *Perspectives on Psychological Science* 5:22–32.

Premack, D., and A. J. Premack. 1994. Levels of causal understanding in chimpanzees and children. *Cognition* 50: 347–62.

Premack, D., and G. Woodruff. 1978. Does the chimpanzee have a theory of mind? *Behavioral and Brain Sciences* 4:515–26.

Preston, S. D. 2013. The origins of altruism in offspring care. *Psychological Bulletin* 139:1305–41.

Price, T. 2013. *Vocal Communication within the Genus* Chlorocebus: *Insights into Mechanisms of Call Production and Call Perception*. Unpublished thesis, University of Göttingen, Germany.

Prior, H., A. Schwarz, and O. Güntürkün. 2008. Mirror-induced behavior in the magpie (*Pica pica*): Evidence of self-recognition. *Plos Biology* 6:e202.

Proctor, D., R. A. Williamson, F. B. M. de Waal, and S. F. Brosnan. 2013. Chimpanzees play the ultimatum game. *Proceedings of the National Academy of Sciences USA* 110: 2070–75.

Proust, M. 1913–27. *Remembrance of Things Past*, vol. 1, *Swann's Way and Within a Budding Grove*. New York: Vintage Press.

Pruetz, J. D., and P. Bertolani. 2007. Savanna chimpanzees, *Pan troglodytes verus*, hunt with tools. *Current Biology* 17:412–17.

Raby, C. R., D. M. Alexis, A. Dickinson, and N. S. Clayton. 2007. Planning for the future by western scrub-jays. *Nature* 445:919–21.

Rajala, A. Z., K. R. Reininger, K. M. Lancaster, and L. C. Populin. 2010. Rhesus monkeys (*Macaca mulatta*) do recognize themselves in the mirror: Implications for the evolution of self-recognition. *Plos ONE* 5:e12865.

Range, F., L. Horn, Z. Viranyi, and L. Huber. 2008. The absence of reward induces inequity aversion in dogs. *Proceedings of the National Academy of Sciences USA* 106:340–45.

Range, F., and Z. Virányi. 2014. Wolves are better imitators of conspecifics than dogs. *Plos ONE* 9:e86559.

Reiss, D., and L. Marino. 2001. Mirror self-recognition in the bottlenose dolphin: A case of cognitive convergence. *Proceedings of the National Academy of Sciences USA* 98:5937–42.

Roberts, A. I., S.-J. Vick, S. G. B. Roberts, and C. R. Menzel. 2014. Chimpanzees modify intentional gestures to coordinate a search for hidden food. *Nature Communications* 5:3088.

Roberts, W. A. 2012. Evidence for future cognition in animals. *Learning and Motivation* 43:169–80.

Rochat, P. 2003. Five levels of self-awareness as they unfold early in life. *Consciousness and Cognition* 12:717–31.

Röell, R. 1996. *De Wereld van Instinct: Niko Tinbergen en het Ontstaan van de Ethologie in Nederland (1920–1950).* Rotterdam: Erasmus.

Romanes, G. J. 1882. *Animal Intelligence.* London: Kegan, Paul, and Trench.

———. 1884. *Mental Evolution in Animals.* New York: Appleton.

Sacks, O. 1985. *The Man Who Mistook His Wife for a Hat.* London: Picador.

Saito, A., and K. Shinozuka. 2013. Vocal recognition of owners by domestic cats (*Felis catus*). *Animal Cognition* 16:685–90.

Sanz, C. M., C. Schöning, and D. B. Morgan. 2010. Chimpanzees prey on army ants with specialized tool set. *American Journal of Primatology* 72: 17–24.

Sapolsky, R. 2010. Language. May 21, http://bit.ly/1BUEv9L.

Satel, S., and S. O. Lilienfeld. 2013. *Brain Washed: The Seductive Appeal of Mindless Neuroscience.* New York: Basic Books.

Savage-Rumbaugh, S., and R. Lewin. 1994. *Kanzi: The Ape at the Brink of the Human Mind.* New York: Wiley.

Sayigh, L. S., et al. 1999. Individual recognition in wild bottlenose dolphins: A field test using playback experiments. *Animal Behaviour* 57:41–50.

Schel, M. A., et al. 2013. Chimpanzee alarm call production meets key criteria for intentionality. *Plos ONE* 8:e76674.

Schusterman, R. J., C. Reichmuth Kastak, and D. Kastak. 2003. Equivalence classification as an approach to social knowledge: From sea lions to simians. In *Animal Social Complexity*, ed. F. B. M. de Waal and P. L. Tyack, 179–206. Cambridge, MA: Harvard University Press.

Semendeferi, K., A. Lu, N. Schenker, and H. Damasio. 2002. Humans and great apes share a large frontal cortex. *Nature Neuroscience* 5:272–76.

Sheehan, M. J., and E. A. Tibbetts. 2011. Specialized face learning is associated with individual recognition in paper wasps. *Science* 334:1272–75.

Shettleworth, S. J. 1993. Varieties of learning and memory in animals. *Journal of Experimental Psychology: Animal Behavior Processes* 19:5–14.

———. 2007. Planning for breakfast. *Nature* 445:825–26.

———. 2010. Q&A. *Current Biology* 20: R910–11.

———. 2012. *Fundamentals of Comparative Cognition*. Oxford: Oxford University Press.

Siebenaler, J. B., and D. K. Caldwell. 1956. Cooperation among adult dolphins. *Journal of Mammalogy* 37:126–28.

Silberberg, A., and D. Kearns. 2009. Memory for the order of briefly presented numerals in humans as a function of practice. *Animal Cognition* 12:405–7.

Skinner, B. F. 1938. *The Behavior of Organisms*. New York: Appleton-Century-Crofts.

———. 1956. A case history of the scientific method. *American Psychologist* 11:221–33.

———. 1969. *Contingencies of Reinforcement*. New York: Appleton-Century-Crofts.

Slocombe, K., and K. Zuberbühler. 2007. Chimpanzees modify recruitment screams as a function of audience composition. *Proceedings of the National Academy of Sciences USA* 104:17228–33.

Smith, A. 1976 [orig. 1759]. *A Theory of Moral Sentiments*, ed. D. D. Raphael and A. L. Macfie. Oxford: Clarendon.

Smith, J. D., et al. 1995. The uncertain response in the bottlenosed dolphin (*Tursiops truncatus*). *Journal of Experimental Psychology: General* 124:391–408.

Sober, E. 1998. Morgan's canon. In *The Evolution of Mind*, ed. D. D. Cummins and Colin Allen, 224–42. Oxford: Oxford University Press.

Soltis, J., et al. 2014. African elephant alarm calls distinguish between threats from humans and bees. *Plos ONE* 9:e89403.

Sorge, R. E., et al. 2014. Olfactory exposure to males, including men, causes stress and related analgesia in rodents. *Nature Methods* 11:629–32.

Spocter, M. A., et al. 2010. Wernicke's area homologue in chimpanzees (*Pan troglodytes*) and its relation to the appearance of modern human language. *Proceedings of the Royal Society B* 277:2165–74.

St. Amant, R., and T. E. Horton. 2008. Revisiting the definition of animal tool use. *Animal Behaviour* 75:1199–208.

Stenger, V. J. 1999. The anthropic coincidences: A natural explanation. *Skeptical Intelligencer* 3:2–17.

Stix, G. 2014. The "it" factor. *Scientific American*, Sept., pp. 72–79.

Suchak, M., and F. B. M. de Waal. 2012. Monkeys benefit from reciprocity without the cognitive burden. *Proceedings of the National Academy of Sciences USA* 109:15191–96.

Suchak, M., T. M. Eppley, M. W. Campbell, and F. B. M. de Waal. 2014. Ape duos and trios: Spontaneous cooperation with free partner choice in chimpanzees. *PeerJ* 2:e417.

Suddendorf, T. 2013. *The Gap: The Science of What Separates Us from Other Animals.* New York: Basic Books.

Suzuki, T. N. 2014. Communication about predator type by a bird using discrete, graded and combinatorial variation in alarm call. *Animal Behaviour* 87:59–65.

Tan, J., and B. Hare. 2013. Bonobos share with strangers. *Plos ONE* 8:e51922.

Taylor, A. H., et al. 2014. Of babies and birds: Complex tool behaviours are not sufficient for the evolution of the ability to create a novel causal intervention. *Proceedings of the Royal Society B* 281:20140837.

Taylor, A. H., and R. D. Gray. 2009. Animal cognition: Aesop's fable flies from fiction to fact. *Current Biology* 19:R731–32.

Taylor, A. H., G. R. Hunt, J. C. Holzhaider, and R. D. Gray. 2007. Spontaneous metatool use by New Caledonian crows. *Current Biology* 17:1504–7.

Taylor, J. 2009. *Not a Chimp: The Hunt to Find the Genes That Make Us Human.* Oxford: Oxford University Press.

Terrace, H. S., L. A. Petitto, R. J. Sanders, and T. G. Bever. 1979. Can an ape create a sentence? *Science* 206:891–902.

Thomas, R. K. 1998. Lloyd Morgan's Canon. In *Comparative Psychology: A Handbook,* ed. G. Greenberg and M. M. Haraway, 156–63. New York: Garland.

Thompson, J. A. M. 2002. Bonobos of the Lukuru Wildlife Research Project. In *Behavioural Diversity in Chimpanzees and Bonobos,* ed. C. Boesch, G. Hohmann, and L. Marchant, 61–70. Cambridge: Cambridge University Press.

Thompson, R. K. R., and C. L. Contie. 1994. Further reflections on mirror usage by pigeons: Lessons from Winnie-the-Pooh and Pinocchio too. In *Self-Awareness in Animals and Humans,* ed. S. T. Parker et al., 392–409. Cambridge: Cambridge University Press.

Thorndike, E. L. 1898. Animal intelligence: An experimental study of the associate processes in animals. *Psychological Reviews, Monograph Supplement* 2.

Thorpe, W. H. 1979. *The Origins and Rise of Ethology: The Science of the Natural Behaviour of Animals.* London: Heineman.

Tinbergen, N. 1953. *The Herring Gull's World.* London: Collins.

———. 1963. On aims and methods of ethology. *Zeitschrift für Tierpsychologie* 20: 410–40.

Tinbergen, N., and W. Kruyt. 1938. Über die Orientierung des Bienenwolfes (*Philanthus triangulum* Fabr.). III. Die Bevorzugung bestimmter Wegmarken. *Zeitschrift für Vergleichende Physiologie* 25:292–334.

Tinklepaugh, O. L. 1928. An experimental study of representative factors in monkeys. *Journal of Comparative Psychology* 8:197–236.

Toda, K., and S. Watanabe. 2008. Discrimination of moving video images of self by pigeons (*Columba livia*). *Animal Cognition* 11:699–705.

Tolman, E. C. 1927. A behaviorist's definition of consciousness. *Psychological Review* 34:433–39.

Tomasello, M. 2014. *A Natural History of Human Thinking.* Cambridge, MA: Harvard University Press.

———. 2008. Origins of human cooperation. Tanner Lecture, Stanford University, Oct. 29–31.

Tomasello, M., and J. Call. 1997. *Primate Cognition.* New York: Oxford University Press.

Tomasello, M., A. C. Kruger, and H. H. Ratner. 1993. Cultural learning. *Behavioral and Brain Sciences* 16:495–552.

Tomasello, M., E. S. Savage-Rumbaugh, and A. C. Kruger. 1993. Imitative learning of actions on objects by children, chimpanzees, and enculturated chimpanzees. *Child Development* 64:1688–705.

Tramontin, A. D., and E. A. Brenowitz. 2000. Seasonal plasticity in the adult brain. *Trends in Neurosciences* 23:251–58.

Troscianko, J., et al. 2012. Extreme binocular vision and a straight bill facilitate tool use in New Caledonian crows. *Nature Communications* 3:1110.

Tsao, D., S. Moeller, and W. A. Freiwald. 2008. Comparing face patch systems in macaques and humans. *Proceedings of the National Academy of Sciences USA* 105:19514–19.

Tulving, E. 2005. Episodic memory and autonoesis: Uniquely human? In *The Missing Link in Cognition,* ed. H. Terrace and J. Metcalfe, 3–56. Oxford: Oxford University Press.

———. 1972. Episodic and semantic memory. In *Organization of Memory,* ed. E. Tulving and W. Donaldson, 381–403. New York: Academic Press.

———. 2001. Origin of autonoesis in episodic memory. In *The Nature of*

Remembering: Essays in Honor of Robert G. Crowder, ed. H. L. Roediger et al., 17–34. Washington, DC: American Psychological Association.

Uchino, E., and S. Watanabe. 2014. Self-recognition in pigeons revisited. *Journal of the Experimental Analysis of Behavior* 102:327–34.

Udell, M.A.R., N. R. Dorey, and C.D.L. Wynne. 2008. Wolves outperform dogs in following human social cues. *Animal Behaviour* 76:1767–73.

———. 2010. What did domestication do to dogs? A new account of dogs' sensitivity to human actions. *Biological Review* 85:327–45.

Uexküll, J. von. 1909. *Umwelt und Innenwelt der Tiere.* Berlin: Springer.

———. 1957 [orig. 1934]. A stroll through the worlds of animals and men. A picture book of invisible worlds. In *Instinctive Behavior*, ed. C. Schiller, 5–80. London Methuen.

Vail, A. L., A. Manica, and R. Bshary. 2014. Fish choose appropriately when and with whom to collaborate. *Current Biology* 24:R791–93.

van de Waal, E., C. Borgeaud, and A. Whiten. 2013. Potent social learning and conformity shape a wild primate's foraging decisions. *Science* 340:483–85.

van Hooff, J. A. R. A. M. 1972. A comparative approach to the phylogeny of laughter and smiling. In *Non-Verbal Communication,* ed. R. A. Hinde, 209–41. Cambridge: Cambridge University Press.

van Leeuwen, E. J. C., K. A. Cronin, and D. B. M. Haun. 2014. A group-specific arbitrary tradition in chimpanzees (*Pan troglodytes*). *Animal Cognition* 17:1421–25.

van Leeuwen, E. J. C., and D. B. M. Haun. 2013. Conformity in nonhuman primates: Fad or fact? *Evolution and Human Behavior* 34:1–7.

van Schaik, C. P., L. Damerius, and K. Isler. 2013. Wild orangutan males plan and communicate their travel direction one day in advance. *Plos ONE* 8:e74896.

van Schaik, C. P., R. O. Deaner, and M. Y. Merrill. 1999. The conditions for tool use in primates: Implications for the evolution of material culture. *Journal of Human Evolution* 36:719–41.

Varki, A., and D. Brower. 2013. *Denial: Self-Deception, False Beliefs, and the Origins of the Human Mind.* New York: Twelve.

Vasconcelos, M., K. Hollis, E. Nowbahari, and A. Kacelnik. 2012. Pro-sociality without empathy. *Biology Letters* 8:910–12.

Vauclair, J. 1996. *Animal Cognition: An Introduction to Modern Comparative Psychology.* Cambridge, MA: Harvard University Press.

Visalberghi, E., and L. Limongelli. 1994. Lack of comprehension of cause-effect relations in tool-using capuchin monkeys (*Cebus apella*). *Journal of Comparative Psychology* 108:15–22.

Visser, I. N., et al. 2008. Antarctic peninsula killer whales (*Orcinus orca*) hunt seals and a penguin on floating ice. *Marine Mammal Science* 24:225–34.

Wade, N. 2014. *A Troublesome Inheritance: Genes, Race and Human History.* New York: Penguin.

Wallace, A. R. 1869. Sir Charles Lyell on geological climates and the origin of species. *Quarterly Review* 126:359–94.

Wascher, C. A. F., and T. Bugnyar. 2013. Behavioral responses to inequity in reward distribution and working effort in crows and ravens. *Plos ONE* 8:e56885.

Wasserman, E. A. 1993. Comparative cognition: Beginning the second century of the study of animal intelligence. *Psychological Bulletin* 113:211–28.

Watanabe, A., U. Grodzinski, and N. S. Clayton. 2014. Western scrub-jays allocate longer observation time to more valuable information. *Animal Cognition* 17:859–67.

Watson, S. K., et al. 2015. Vocal learning in the functionally referential food grunts of chimpanzees. *Current Biology* 25:1–5.

Weir, A. A., J. Chappell, and A. Kacelnik. 2002. Shaping of hooks in New Caledonian crows. *Science* 297:981–81

Wellman, H. M., A. T. Phillips, and T. Rodriguez. 2000. Young children's understanding of perception, desire, and emotion. *Child Development* 71: 895–912.

Wheeler, B. C., and J. Fischer. 2012. Functionally referential signals: A promising paradigm whose time has passed. *Evolutionary Anthropology* 21:195–205.

White, L. A. 1959. The *Evolution of Culture*. New York: McGraw-Hill.

Whitehead, H., and L. Rendell. 2015. *The Cultural Lives of Whales and Dolphins.* Chicago: University of Chicago Press.

Whiten, A., V. Horner, and F. B. M. de Waal. 2005. Conformity to cultural norms of tool use in chimpanzees. *Nature* 437:737–40.

Wikenheiser, A., and A. D. Redish. 2012. Hippocampal sequences link past, present, and future. *Trends in Cognitive Sciences* 16:361–62.

Wilcox, S., and R. R. Jackson. 2002. Jumping spider tricksters: Deceit, predation, and cognition. In the *Cognitive Animal: Empirical and Theoretical Perspectives on Animal Cognition*, ed. M. Bekoff, C. Allen, and G. Burghardt, 27–33. Cambridge, MA: MIT Press.

Wilfried, E. E. G., and J. Yamagiwa. 2014. Use of tool sets by chimpanzees for multiple purposes in Moukalaba-Doudou National Park, Gabon. *Primates* 55:467–42.

Wilson, E. O. 1975. *Sociobiology: The New Synthesis*. Cambridge, MA: Belknap Press.

———. 2010. *Anthill: A Novel*. New York: Norton.

Wilson, M. L., et al. 2014. Lethal aggression in *Pan* is better explained by adaptive strategies than human impacts. *Nature* 513:414–17.

Wittgenstein, L. 1958 [orig. 1953]. *Philosophical Investigations*, 2nd ed. Oxford: Blackwell.

Wohlgemuth, S., I. Adam, and C. Scharff. 2014. FOXP2 in songbirds. *Current Opinion in Neurobiology* 28:86–93.

Wynne, C. D., and M. A. R. Udell. 2013. *Animal Cognition: Evolution, Behavior and Cognition*. 2nd. ed. New York: Palgrave Macmillan.

Yamakoshi, G. 1998. Dietary responses to fruit scarcity of wild chimpanzees at Bossou, Guinea: Possible implications for ecological importance of tool use. *American Journal of Physical Anthropology* 106:283–95.

Yamamoto, S., T. Humle and M. Tanaka. 2009. Chimpanzees help each other upon request. *Plos One* 4:e7416.

Yerkes, R. M. 1925. *Almost Human*. New York: Century.

———. 1943. *Chimpanzees: A Laboratory Colony*. New Haven, CT: Yale University Press.

Zahn-Waxler, C., M. Radke-Yarrow, E. Wagner, and M. Chapman. 1992. Development of concern for others. *Developmental Psychology* 28:126–36.

Zylinski, S. 2015. Fun and play in invertebrates. *Current Biology* 25:R10–12.

GLOSSARY

Analogy: Structurally and functionally similar traits (such as the streamlined shapes of fish and dolphins) that evolved independently as adaptations to the same environment. See also: convergent evolution.

Anthropocentrism: A worldview revolving around the human species.

Anthropodenial: The a priori denial of humanlike characteristics in other animals or animallike characteristics in humans.

Anthropomorphism: The (mis)attribution of humanlike characteristics and experiences to other species.

Behaviorism: The psychological approach introduced by B. F. Skinner and John Watson with emphasis on observable behavior and learning. In its more extreme form, behaviorism reduces behavior to learned associations and rejects internal cognitive processes.

Biologically prepared learning: Learning talents and predispositions that evolved to suit a species's ecology and assist its survival. See also: Garcia Effect.

Bonding- and Identification-based Observational Learning (BIOL): Social learning primarily based on a desire to belong and conform to social models.

Clever Hans Effect: The influence of unintentional cues by the experimenter in inducing an apparent cognitive feat.

Cognition: The transformation of sensory input into knowledge about the environment, and the application of this knowledge.

Cognitive ethology: Donald Griffin's label for the biological study of cognition.

Cognitive ripple rule: The rule that every cognitive capacity turns out to be older and more widespread than initially assumed.

Comparative psychology: The subdiscipline of psychology that seeks to find general principles of animal and human behavior or, more narrowly, to use animals as models for human learning and psychology.

Conformist bias: The tendency of an individual to favor the solutions and preferences of the majority.

Conspecific approach: The technique of testing animals with models or partners of their own species in order to reduce human influence.

Convergent evolution: The independent evolution of similar traits or capacities in unrelated species in response to similar environmental pressures. See also: analogy.

Cooperative pulling paradigm: An experimental paradigm in which two or more individuals pull rewards toward themselves via an apparatus that they cannot successfully operate alone.

Critical anthropomorphism: The use of human intuitions about a species to generate objectively testable ideas.

Culture: The learning of habits and traditions from others, with the result that groups of the same species behave differently.

Delayed gratification: The ability to resist an immediate reward in order to receive a better one later on.

Displacement activity: An activity irrelevant to the current situation that appears suddenly due to a thwarted motivation or conflict between incompatible motivations, such as fight and flight.

Ecological niche: The role of a species within an ecosystem and the natural resources it relies on.

Episodic memory: The recollection of specific past experiences, such as their content, location, and timing.

Embodied cognition: A view of cognition that emphasizes the role of the body (beyond the brain) and its interaction with the environment.

Ethology: The biological approach to animal and human behavior, introduced by Konrad Lorenz and Niko Tinbergen, that emphasizes species-typical behavior as an adaptation to the natural environment.

Evolutionary cognition: The study of all cognition, human and animal, from an evolutionary perspective.

Function: The purpose of a trait as measured by the benefits it confers.

Garcia Effect: An aversion to a specific food develops after negative effects, such as nausea and vomiting, even if these effects occurred after a long time interval. See also: biologically prepared learning.

Homology: A similarity in traits of two species that is explained by the presence of these traits in their common ancestor.

Hume's Touchstone: David Hume's plea to apply the same hypotheses to the mental operations of both humans and animals.

Inferential reasoning: The use of available information to construct a reality that is not directly observable.

Insight: The sudden combining (aha! experience) of past bits of information to mentally come up with a novel solution to a novel problem.

Intelligence: The ability to successfully apply information and cognition to solve problems.

Killjoy account: The deflation of a claim regarding higher mental processes by proposing a seemingly simpler explanation.

Know-thy-animal rule: The rule that anyone who questions a cognitive claim about a species should either be familiar with the species or make an effort to verify the counterclaim.

Magic well: The endless complexity of the specialized cognition of any organism.

Matching-to-sample paradigm: An experimental framework in which the subject, after perceiving a sample, must find a matching one from among two or more options.

Mental time travel: An individual's awareness of its own past and future.

Metacognition: The monitoring of one's own memory in order to know what one knows.

Mirror mark test: An experiment to determine whether an organism will notice a mark on its body that it can see only via its mirror image.

Morgan's Canon: The advice not to assume higher cognitive capacities if lower ones may explain an observed phenomenon.

Object permanence: The realization that an object continues to exist even after it has disappeared from an individual's perception.

Overimitation: The imitation of all the actions shown by a model even if not all serve to reach the goal.

Perspective taking: The ability to look at a situation from another's standpoint.

Pygmalion Effect: The way a given species is tested often reflects cognitive prejudices. Specifically: comparative testing favors our own species.

Scala naturae: The natural scale of the ancient Greeks ranking all organisms from low to high, with humans closest to the angels.

Selective imitation: Imitation of only those actions that lead to the goal while ignoring other behavior.

Self-awareness: Consciousness of the self, which some interpret as requiring an organism to pass the mirror mark test, whereas others believe that self-awareness characterizes all life-forms.

Signature whistles: Dolphin calls modulated so that each individual has a distinct and recognizable "melody."

Social brain hypothesis: The hypothesis that the relatively large brain size of primates is explained by the complexity of their societies and their need to process social information.

Targeted helping: Assistance given by one individual to another based on perspective taking, such as judgment of the other's specific situation and needs.

Theory of mind: The ability to attribute mental states to others, such as knowledge, intentions, and beliefs.

Triadic awareness: Individual A's knowledge not only of its own relations with individuals B and C, but also of the relation between B and C themselves.

True imitation: A subtype of imitation that reflects understanding of the other's methods and goals.

Umwelt: An organism's subjective perceptual world.

ACKNOWLEDGMENTS

My interest in cognition as an evolved characteristic marks me as an ethologist. I am grateful to all the Dutch ethologists, who influenced my early career. I began my graduate studies at the University of Groningen, in the Netherlands, under Gerard Baerends, who was Niko Tinbergen's first student. Afterward I wrote my dissertation on primate behavior at the University of Utrecht with Jan van Hooff, an expert of facial expressions and emotions. My exposure to comparative psychology, the other approach to animal behavior, came mostly after my move across the Atlantic. Input from both schools has been critical to constructing the new field of evolutionary cognition. This book relates my own journey and involvement in this field as it gradually moved to the forefront of the study of animal behavior.

I am grateful to the many people who have accompanied me on this journey, from colleagues and collaborators to students and postdocs. Just to thank those of the last few years: Sarah Brosnan, Kimberly Burke, Sarah Calcutt, Matthew Campbell, Devyn Carter, Zanna Clay, Marietta Danforth, Tim and Katie Eppley, Pier Francesco Ferrari, Yuko Hattori, Victoria Horner, Joshua Plotnik, Stephanie Preston, Darby Proctor, Teresa Romero, Malini Suchak, Julia Watzek, Christine Webb, and Andrew Whiten. I am grateful to the Yerkes National Primate Research Center and Emory University for the opportunity to conduct our studies, and to the many monkeys and apes who have participated and become part of my life.

This book was initially undertaken as a relatively short overview of recent findings in primate cognition but quickly grew in scope and size to what it is now. Inclusion of other species has been paramount, because the field of animal cognition has become much more varied in the last two decades. This overview is obviously incomplete, but my main objective is to convey enthusiasm for evolutionary cognition and to illustrate how it has grown into a respectable science based on rigorous observations and experiments. Since the book covers so many different aspects and species, I have asked colleagues to read parts of it. For their invaluable feedback, I thank Michael Beran, Gregory Berns, Redouan Bshary, Zanna Clay, Harold Gouzoules, Russell Gray, Roger Hanlon, Robert Hampton, Vincent Janik, Karline Janmaat, Gema Martin-Ordas, Gerald Massey, Jennifer Mather, Tetsuro Matsuzawa, Caitlin O'Connell, Irene Pepperberg, Bonnie Perdue, Susan Perry, Joshua Plotnik, Rebecca Snyder, and Malini Suchak.

I thank my agent Michelle Tessler for her continued support, and my editor at Norton, John Glusman, for critical reading of the manuscript. As always, my wife and number-one fan, Catherine, has read my daily production with enthusiasm and helped me stylistically. I thank her for the love in my life.

INDEX

Note: Page numbers in *italics* refer to illustrations

AUTHOR BIOGRAPHY

Frans de Waal is a Dutch-American ethologist and primatologist. Having earned a Ph.D. in biology from the University of Utrecht, in 1977, he completed a six-year study of the chimpanzee colony at Burgers' Zoo in Arnhem, before moving to the United States. His first popular book, *Chimpanzee Politics*, compared the schmoozing and scheming of chimpanzees involved in power struggles with that of human politicians. Ever since, de Waal has drawn parallels between primate and human behavior. Translated into over twenty languages, his books have made him one of the world's most visible biologists.

With his discovery of reconciliation in primates, de Waal pioneered research on animal conflict resolution. He received the 1989 Los Angeles Times Book Award for *Peacemaking among Primates*. His scientific articles have been published in journals from *Science*, *Nature*, and *Scientific American* to those specialized in animal behavior and cognition. His latest interests are animal cooperation, emotion, and empathy, and the evolution of human morality.

De Waal is C. H. Candler Professor in the Psychology Department of Emory University; Director of the Living Links Center at the Yerkes National Primate Research Center, in Atlanta; and Distinguished Professor at the University of Utrecht. He is a longtime member of the board of directors of Chimp Haven, the National Chimpanzee Sanctuary, which releases ex-laboratory chimpanzees on large forested islands in Louisiana. He has been elected to both the U.S. National Academy of

Sciences and the Royal Netherlands Academy of Arts and Sciences. In 2007 he was selected by *Time* as one of the World's 100 Most Influential People Today and in 2011 by *Discover* as among forty-seven (all-time) Great Minds of Science.

With his wife, Catherine, de Waal lives in Smoke Rise, Georgia.